INJECTION MOULDING
OF
ELASTOMERS

INJECTION MOULDING

OF ELASTOMERS

Edited by

W. S. Penn B.SC., A.I.R.I.

MACLAREN AND SONS

LONDON

85334 054 4
INJECTION MOULDING
OF ELASTOMERS
EDITED BY W. S. PENN
FIRST PUBLISHED 1969 BY
MACLAREN AND SONS LTD
7 GRAPE STREET
LONDON WC2
COPYRIGHT © 1969 BY
MACLAREN AND SONS LTD

COMPOSED AND PRINTED BY
PERIDON LTD., LONDON, ENGLAND

Editors' preface

THE use of injection moulding in the rubber industry has reached an unusual position. Whilst appreciating the advantages of the process in certain instances, the industry, so often conservative, has been hovering on the brink of enthusiastic acceptance for many years. Apart from any technical aspects this is probably due to the relatively high cost of capital equipment and teething troubles with existing machines.

The time thus seemed appropriate to appraise the situation to assist rubber manufacturers in making their choice. A conference was, therefore, organised at the Borough Polytechnic to examine all aspects of the subject. In the event it was highly successful and many useful discussions took place. This book is a record of the papers and the discussions which followed them.

It should be stressed that although the book is a record of the proceedings, the lecturers and subjects were chosen with care to present a complete and logical study of the subject. In this sense, therefore, the publication may be regarded as a textbook on this relatively new subject and virtually all aspects are considered in some depth.

Thus the book is broadly divided into three parts. The first deals with general principles, engineering aspects and machines; the second deals with compounding elastomers for injection moulding and the third with moulding various elastomers. Economic aspects are considered at all points.

The lectures given in the Conference were composed by authors who had no contact with one another. Inevitably, therefore, there was a certain amount of overlapping, particularly in describing the operation of a screw preplasticizer injection moulding machine and the advantages of the injection moulding of elastomers. Thus, in order to avoid repetition a certain amount of editing has taken place to overcome this problem. Nevertheless, there is still some repetition in those papers where to remove the appropriate parts would have left them out of balance. A particularly full paper, with numerous references on the advantages of injection moulding is that by Dr M. A. Wheelans on 'The injection moulding of natural rubber'.

I must acknowledge first of all the assistance of Mr L. R. Mernagh, Director of the Institution of the Rubber Industry for opening the Conference, being chairman on the first day and giving advice and encouragement. It would be invidious to mention individual lecturers by name but I am extemely grateful to them all for the trouble they went to and the high quality of their lectures: their names are all given in the text. I must also thank the Borough Polytechnic for providing such excellent facilities and my colleagues for their help.

Finally, I must thank the publishers for their assistance and handling of a very difficult manuscript.

August 1968   W. S. PENN

The Conference on Injection Moulding of Elastomers was held at

The Borough Polytechnic, Borough Road, London, SE1

on March 12, 13 and 14, 1968

Contents

List of plates

Introduction

L. R. MERNAGH
Director, The Institution of the Rubber Industry

THE Borough Polytechnic has long been renowned as a seat of learning for Plastics Technology. In fact it has recently been approved by the Inner London Education Authority as one of two Polymer Technology Institutions for London, the other being the National College, Holloway. Recently 'The Boro' has decided to spread its wings and take in Rubber Technology—an event not totally dissociated from the appointment of Mr W. S. Penn to its staff. Mr Penn, who organized this Symposium very efficiently, received his early training in the tyre industry and it was a bold move to confine a Conference on Injection moulding to elastomers. Devoting three days and fifteen lectures to such a subject obviously opened up the possibility of both sales talks and repetitition. These, however, were skilfully minimized and the eighty delegates plus students who attended were rewarded with a mass of useful information.

The injection moulding machine differs from other equipment in our industry in that it is a recent development and applicable to both rubber and plastics and so makes a contribution to bridging the gap between these two. On the other hand, it does accentuate the difference between rubber and plastics. Specific problems with the former include that of vulcanization and, particularly, prevulcanization before the material reaches the mould. There is, in addition, the problem of getting the raw material in a suitable condition to permit hopper feeding. Injection moulding depends for its efficiency on long runs and this is another area in which the rubber industry does not enjoy the same freedom as plastics. Multi-station moulding has helped in this connection but, even if different articles can be produced simultaneously by this means they have to be made from the same compound.

This development has probably suffered because initially the proposition was oversold and injection moulding machines were claimed to perform well beyond their ability. The value of such a Conference as this is to bring the whole subject into the right perspective, and it is obvious that a great deal needs to be done to equate rubber processing to that of plastics.

In the pages that follow it will be seen that the stage was very well set by an introductory talk by Mr Izod of RAPRA on the principles of injection moulding. This was followed by a description of an impressive array, both of equipment and materials, with a very clear indication that there is considerable cooperation between manufacturers and users. Most of the polymers, both wellknown and speciality, are dealt with including their compounding. The special requirements for mixed stocks for injection moulding include good flow characteristics with rapid vulcanization at low temperature with freedom from pre-cure at the same time. These necessitate good delayed action accelerators, and again draw attention to the EV system of vulcanization.

The names of many well-known authorities on injection moulding and raw materials appear in the list of speakers, and active discussion followed most of their papers. This is adequately summarized in the pages which follow.

L.R.M.

CHAPTER 1

Some fundamental aspects of injection moulding of elastometers

by D. A. W. IZOD and G. D. SKAM
Rubber and Plastics Research Association

1.1 INTRODUCTION

THE controllable variables in injection moulding are injection pressure, barrel temperature, screw speed and mould temperature of screw ram machines; and the injection pressure, blank temperature and mould temperature for piston types. These variables and their effects are examined.

1.2 INJECTION PRESSURE

The importance of injection pressure is its effect on heat generation although if pressure is allowed to fall below a certain level it will have an effect on injection time. This critical level is influenced by choice of polymer and by nozzle size, assuming uniform preheating of the compound to working level. This effect is shown in Fig. 1.1.
These curves were obtained using a Turner CTA-2-80S machine to inject stocks based on different polymers with 50 phr loadings of HAF black. The nozzle size was chosen to give reasonable injection times at a realistic working pressure. It shows clearly the ease with which cis-polyisoprene can be injected.

Although acceptably short injection times can often be obtained with a large nozzle, this will result in loss of heat build-up. On most machines the hydraulic system requires a finite time to reach its maximum rated pressure. Hence if a very large nozzle is used to give a short injection time, injection may well be complete before maximum pressure can be utilized; this will result in lower temperature rise during injection.

The temperature rise is related directly to the work performed during injection and may be calculated from

$$\Delta t = \frac{I\,g}{J\,d\,s \times 10^7}$$

where

I = injection pressure
d = specific gravity of compound
s = specific heat of compound
J = mechanical equivalent of heat
g = acceleration due to gravity.

Thus for an injection pressure of 1 480 kg./cm.^{2} (21 000 psi) a theoretical temperature rise of 60°C would be expected,

1

(when d = 1·15 and S = 0·5) approximating to 3°C/1 000 psi injection pressure.

In practice, owing to dynamic losses in the system and possible failure of the pump to reach its maximum working pressure, lower values will be obtained. Fig. 1.2 shows graphically experimental data obtained with a Foster Yates and Thom 'Rubberometer', the line having a slope of 2·3°C/1 000 psi. Subsequent work with the Turner CTA-2-80S machine supports these calculations.

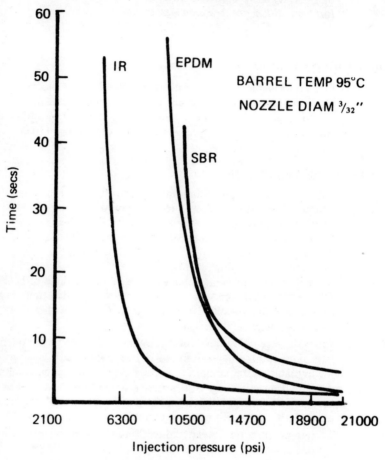

Fig. 1.1 Injection pressure vs *injection time.*

1.3 BARREL TEMPERATURE

In order to minimize cure times, the stock must be injected at a temperature as close as safety permits to that of the moulding cavity. This temperature is obtained partly by the heat generated during injection, but primarily by the temperature of the compound before injection. This is determined by the barrel temperature and the work done on the compound by the screw.

The reduction of cure time obtained by raising stock temperature is noticeable (with screw ram machines) even with thin mouldings and a 20°C increase in barrel temperature, as shown in Fig. 1.3.

2

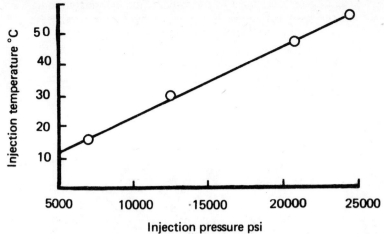

Fig. 1.2. Injection pressure vs injection temperature data from
Foster Yates and Thom 'Rubberometer'

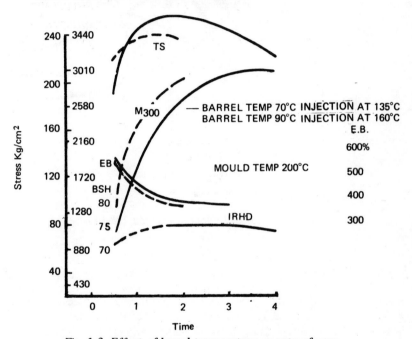

Fig. 1.3. Effect of barrel temperature on rate of cure.

1.4 OPTIMIZING OF MACHINE CONDITIONS

A 'thermal load' diagram of the type shown in Fig. 1.4 after the
manner of Kleine-Albers and Franck of Ankawerk, indicates the
technique to be followed in optimizing machine variables. The necessary
increase in stock temperature is achieved by adjustment of the barrel
temperature, and the diagram shows plainly the advantages to be
obtained, and this advantage increases with increasing moulding
thickness. Temperature rise through the nozzle can be experimentally
obtained by inserting a needle pyrometer in the rubber before injection,

3

and again after injecting the material into an insulated container. For a temperature rise of approximately 35°C and an injection temperature 20°C below the mould temperature (180°C) the temperature to be set for the rubber in the barrel is 125°C. The compound must, of course, be safe from scorching at temperatures reached in the barrel. A useful rule for barrel temperatures of 90–120°C is a Mooney scorch time (at 121°C) of more than twice the dwell time of the rubber in the screw flights.

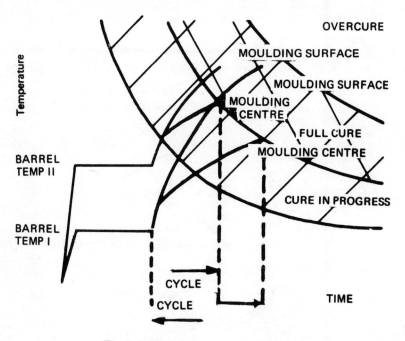

Fig. 1.4. Optimizing screw ram machines.

Further examination of Fig. 1.4 reveals that another advantage to be obtained from the use of higher barrel temperatures is the reduction in cure differentials between sections of different thickness in the same moulding. This is shown by the reduced difference between the attained temperatures of surface and centre in the two cycles.

A similar thermal load diagram for piston machines (Fig. 1.5) shows the advantages to be gained from the use of blank preheating. Many of the piston type machines now being offered for sale in Europe have no built in provision for preheating the blank. This has little effect on cure time if the moulding thickness is small, even up to ½–¾ in., although it will still be beneficial in reducing injection time, which may otherwise be long with a cold blank. With thicker mouldings reductions in cure time of about 50% have been reported when using micro-wave heaters. Micro-wave heating is probably the most efficient method of blank preheating but even a simple oven is effective, although in this case the blank must be fed in thin pieces.

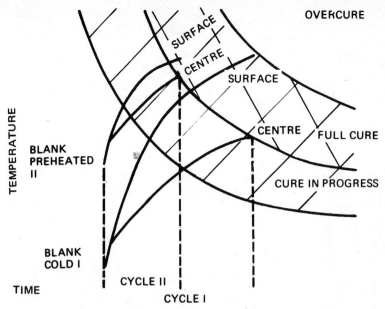

Fig. 1.5. Optimizing piston machines.

1.5 RHEOLOGY OF POLYMERS

The relationship between injection pressure and injection time has already been described, while a further examination of Fig. 1.1 shows how closely the injection times for compounds of different types are grouped together once the pressure has passed a critical point.

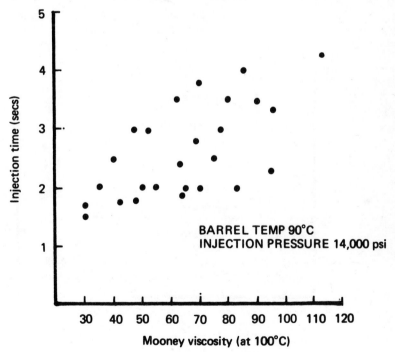

Fig. 1.6. Injection time vs Mooney viscosity.

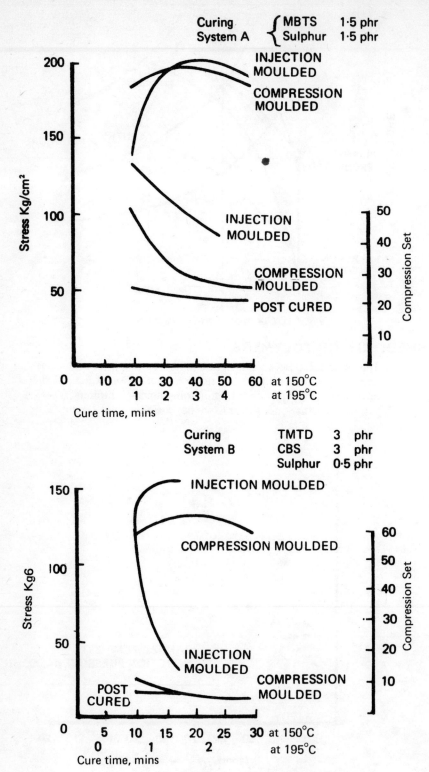

Fig. 1.7. Compression set of nitrile stocks.

This close grouping emphasizes the impossibility of predicting injection behaviour in a screw ram machine from Mooney plasticity figures. Figure 1.6 shows the scatter of results for a variety of Mooney levels; all the injection times are under 5 sec. for Mooney levels ranging between 30 and 115.

The Mooney viscosity is, however, of importance in forecasting behaviour of the stock during preplasticizing in the screw since this corresponds closely to extrusion, and reference has already been made to the importance of Mooney scorch.

1.6 MECHANICAL PROPERTIES OF POLYMERS

It has been reported that injection moulding gives higher tensile strength than does compression moulding, an effect which has been attributed to improved dispersion.

Other property levels may be impaired, however, and reports from some quarters have indicated that compression set is a property likely to be affected in this way.

Using nitrile rubber compounds based on various curing systems, specimens 0·1 in. thick were injection moulded at 195°C. Tensile strength and compression set measurements (the latter using plied up specimens) made on these mouldings are shown in Figs 1.7 a and b.

These show that although the tensile strength of the injection moulded specimen is higher than that of the compression moulded, the compression set of the injection moulded specimens is also markedly higher. However, when a post-cure of 60 min. at 150°C is given, the compression set of both sets of mouldings is comparable.

Tensile strength measurements made on similar compounds which have been press cured at 195°C indicate that the impaired results are attributable to the higher temperature of cure rather than any feature of injection moulding as such. Using the very simple MBTS system, the compression set has reached 45% after 5 min. cure, and when the efficient TMTD/CBS/low sulphur system is employed a figure of 15% is reached. When a post cure of 60 min. at 150°C is given the results are again comparable with those obtained on specimens cured at conventional press temperatures.

The increased use of injection moulding is now accompanied by a better understanding of the basic factors involved. This allows the process to be controlled as readily as the older long-established processes. A sufficiently accurate prediction of the temperature rise on injection is possible, and a simple method of optimizing conditions can be applied.

REFERENCES

1 This paper has been published by RAPRA as Research Report 167.
2 Previous work on the subject by RAPRA is given in Research Report 139.

CHAPTER 2

Injection moulding and its comparison with other methods

H. M. GARDNER
Daniels of Stroud Ltd

2.1 INTRODUCTION

INJECTION moulding is a process which has been used for many years by both the rubber and plastics industries although it is only in recent times that the specialized equipment available has been accepted in any volume by the rubber industry. Indeed the number of rubber moulders using such equipment is still quite small, and many are still quite sceptical of the economies of the process. The purpose of this paper is to illustrate some of the types of machine available and to outline the situations where their use can give better overall results than conventional methods.

More specifically, this paper will deal with plunger type transfer moulding presses, and single-screw injection moulding machines; it will not cover the use of transfer moulds incorporating an integral transfer chamber using a simple compression moulding press to provide both the transfer and mould locking force.

2.2 TRANSFER MOULDING

It is convenient to consider the transfer moulding press first, as this is the simplest process to describe and the screw injection machine utilizes similar techniques with certain refinements. Figure 2.1 Plate 1 shows a typical example of such a machine. It comprises a fast operating downstroke press which serves to lock the mould during injection and cure, and an upstroking transfer or injection unit. A combination of upstroking press and downstroking transfer unit could equally well be used.

Machines of this type are usually built as self-contained units, incorporating a hydraulic pump and motor, and a cycle controller on which the moulding sequence can be preset. This latter feature is very necessary on these machines. The cure time is short and the minor differences from cycle to cycle which would be present under conditions of manual control are best eliminated. Cure times can be as short as a minute, and errors of a few seconds can have a noticeable effect.

Transfer presses are, of course, made in a range of sizes, but, for reasons which will be mentioned later, the smallest machine of interest to the

Photographic illustrations for this chapter will be found on Plate 1 (between pages 14-15).

general mechanical goods moulder is one of approximately 100 tons capacity. This will have a capacity of about 1½ lb. of rubber and the platens will be about 18–20 in. square. The transfer unit would be one exerting a force of 40–50 tons.

It is much more difficult to define the larger machine. At the present time there are quite a number of presses of 300–500 tons, with a shot capacity of 12–15 lb., and there would seem to be no reason why future developments should not produce larger units. Indeed, the economies of the process should be much greater with heavier mouldings of thicker section, the only drawback being that such mouldings are not always required in sufficiently large numbers to justify the expense of tooling.

As far as the press in Fig. 2.1 Plate 1 is concerned it has a maximum locking force of 200 tons, and is fitted with platens 26 in. square. These platens are electrically heated, the temperature being controlled by thermostats, temperatures of up to 200°C are usual. The closing speed of the press is about 6 in. per sec., and a cushioning device is provided to reduce the impact of mould closing. The transfer cylinder has a maximum force of 60 tons and should be capable of a speed of up to 1 in. per sec. when injecting the compound, although in practice it will be found that a speed of ½in./sec. will usually be adequate.
Again, it will be found essential to be able to control this speed with some accuracy, as the optimum figure will vary from mould to mould. The time cycle controller will initiate the transfer movement when locking pressure has been built up, open the press at the end of the cure period, and control any ancillary movements such as ejection, core extraction, etc.

The operation of the press is quite simple. The mould consists of one or more cavities, connected to a central transfer chamber by a series of runners. This chamber is loaded by the operator at the beginning of each cycle, using a slug of compound slightly heavier than the weight of the mouldings plus the runner system. Only one slug is required, regardless of the number of cavities. The press is closed and the transfer plunger rises in order to force the rubber into the mould cavities. A reasonably high injection force is used, in the order of 4–6 tons/sq. in. Altough the mould construction may be, and indeed should be, of a high standard with mating surfaces accurately matched, the force available will be sufficient to cause any excess rubber to be pushed through the cavity to form flash along the parting line.
Such flash is undesirable, particularly as it will vary according to the weight of the slug loaded into the press.

This can be avoided by arranging that, after the mould cavities have been filled, the injection pressure is reduced to a figure sufficient to maintain consolidation during the cure period but insufficient to give flash. In practice, a figure of 1½–2 tons/sq. in. appears to be suitable. The difficulty is in deciding at what point in the injection stroke this should occur. If the rubber blanks are accurately weighed, this reduction in pressure will occur at precisely the same point, but the press designer cannot make this assumption, and must accept the fact that there will be some variation between charges. This is allowed for in the machine illustrated in Fig. 2.2.

9

When the rubber blank is placed in the transfer chamber it will only partially fill it. As a result, there will be little resistance to the upward movement of the transfer plunger. However, once the rubber has been consolidated in the transfer chamber and runner system, a considerably greater pressure will be required in order to force the rubber through the gates, which are normally of small cross-section. This increase in pressure is detected and used to initiate a metering device. From this point onwards the movement of the transfer plunger is measured, and after a preset movement has taken place the pressure is reduced. As the metering begins when rubber starts to enter the cavities, rather than at the beginning of the transfer stroke, any excess of rubber remains in the transfer chamber rather than being washed through the mould to form additional flash. In this connection the author is not suggesting that this method eliminates flash. This is possible under favourable conditions, but one should normally be satisfied that the amount of flash is consistent, and that by careful mould design the thickness of flash can be such as to facilitate subsequent finishing operations.

At the end of the cure period the press opens automatically, and, in some cases the moulding will be partially ejected, and then removed by the operator prior to recharging the press and the initiation of the next cycle.

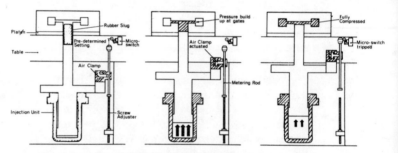

Fig. 2.2 Daniels 'Meterjet' system.

There is a refinement which can be added to the machine if mouldings incorporating metal inserts are being made. A misplaced insert can cause considerable damage if the mould is allowed to close on it and build up pressure. The press can be provided with a device which will detect the presence of an insert or any other undesirable object, and cause the press to open rather than build up pressure. Usually this device also initiates a visual warning so that the mishap can be rectified.

The system just described is perhaps the simplest of injection moulding methods, but it is already obvious that the amount of operator's attention required is relatively small. To take full advantage of the system an operator should be responsible for more than one machine. Reference has already been made to time cycles as short as one minute; this applies to mouldings of thin section. With thicker sections of ¾-1 in. the cycle time is likely to be in the order of 5-7 min., and the operator can reasonably be expected to carry out finishing operations also. When the finish obtained from a tear off flash is acceptable, it is possible to eliminate any additional finishing cost, without accepting the limitations of flash free moulding.

At this stage it is proposed to consider what economies the process has been shown to offer. Blank preparation has been simplified. One blank will suffice, regardless of the number of cavities. The blank is of simple cylindrical shape, and does not have to be preshaped to allow for intricate cavity shape, core pins, etc. Finishing operations can be simplified, or even reduced to the extent that they can be carried out by the press operator.

Although these savings are real, and in some cases quite considerable, they would hardly be sufficient in themselves if the method did not offer a reduction in cure time as well. Fortunately this does occur and is due to two factors. In the first place, owing to the fast operation of the press, higher mould face temperatures can be used without the risk of precure. This in itself will shorten the cure time, but an additional saving occurs owing to the pressure required to force the compound through the restriction presented by the gates. The friction generated in the gate area is sufficient to cause a useful increase in the temperature of the compound, and this occurs throughout the mass of the rubber and not just on the surface, where it is in contact with the mould. As a result, the higher mould temperatures which are used can be tolerated without the risk of scorch on the surface before the centre of the moulding is cured.

Using the system as just described it is reasonable to expect cure times in the order of 25% of those which one would expect from a conventional compression moulding system, using the temperatures that can be obtained from a steam heated mould. With many compounds a further reduction can be obtained by preheating the rubber before it is fed into the press. This is commonly achieved by soaking the rubber blank in hot air for about 30 min. to give an even temperature of 70–80°C throughout the blank. If this method is used, it will be necessary to have six or eight blanks in the preheating device, and, to achieve consistent results care must be taken to ensure that the blanks are used in the correct order, and a new cold blank added each time a hot one is removed to feed the press. Complications can occur if for any reason the press cycle is interrupted for a period. A more efficient method is the use of a micro-wave preheating oven. By this method, not only is the blank heated evenly throughout its thickness, but the time required for heating is appreciably less than the cure time, so that only one blank is being heated. When used with a modern, fast acting press, a preheat temperature of up to 190–200°C can be used with a further reduction in cure time.

An additional advantage is often obtained when preheating is used. The number of mould cavities is frequently limited by the fact that if a certain area is exceeded the press will be forced open by the injection pressure, giving rise to excessive flash. Preheated rubber can usually be injected at a considerably reduced injection pressure owing to its much lower viscosity.

To give some idea of the cure times involved an example may be quoted. The moulding concerned was a natural rubber compound, the section about $\frac{3}{16}$ in. and the compression cure time 7 min. Using a transfer moulding press the cure time was reduced to 2 min., while, when micro-wave preheating was adopted, it was reduced to 70 sec., and owing to the lower injection pressure required, control of flash was very much facilitated.

2.3 INJECTION MOULDING

The single, reciprocating screw injection moulding machine incorporates the features of the transfer moulding press, but is more highly automated. Such a machine is shown in Fig. 2.3 Plate 1. In this case it is a machine of 4-5 oz. capacity. In the plastics industry machines of 300-400 oz. are becoming reasonably common; the rubber industry, in this country, uses machines up to about 30 oz. capacity only.

The sequence of operations is shown diagramatically in Fig. 2.4

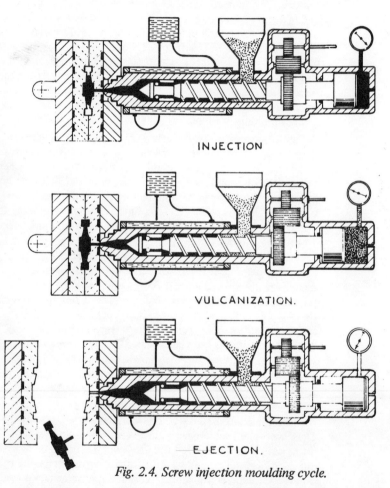

INJECTION

VULCANIZATION.

EJECTION.

Fig. 2.4. Screw injection moulding cycle.

The first diagram shows the mould having been filled, with the injection screw in its forward position. After a few seconds have been allowed for cure to commence, the screw is rotated, causing cold compound from the hopper to be fed forward to the front of the barrel. the barrel is heated to a temperature of about 80°C the rubber has picked up a useful amount of heat by the time it has passed through the screw. At this stage the mould is full, and the rubber cannot pass through the nozzle. As a result it displaces the screw towards the back of the barrel. This backward movement can be measured and used to control the volume of compound available for the next shot. This is one advantage over the transfer moulding press; there is no excess slug of rubber to be wasted.

When the cure time is over, the mould opens and the mouldings are either ejected automatically or removed by the operator. By this time the screw will have fed the next charge to the front of the barrel and stopped rotating. The moulds now close again and the screw is forced forward. It does not rotate, but merely acts as a plunger, forcing rubber into the mould at a pressure of up to 20 000 psi. At this stage the rubber is in an extremely fluid condition, and the mould would flash heavily if subject to the high injection pressure once filled. The pressure is therefore automatically reduced once the injection screw has swept a sufficient volume to fill the mould. After a suitable pause the barrel is recharged and the cycle is repeated.

This description of the injection moulding machine has been deliberately brief as more detailed considerations are given in Chapters 4 and 5.

For the moment, however, it is useful to point out the advantages the system offers compared to transfer moulding:

(1)   Material, in either strip or granular form is fed into the machine automatically.

(2)   Preheating is an inherent feature of the machine, consequently shorter cyle times can be expected.

(3)   The machine is capable of a fully automatic cycle. If the moulding is of a type suitable for automatic ejection, there is no need to have an operator standing by permanently. Unfortunately, few rubber mouldings fall into this category.

In view of these advantages, it is reasonable to ask why the transfer moulding press should be considered at all. The answer is a question of cost. It was mentioned earlier that very small transfer presses are seldom used. The smallest injection moulding machine such as the one illustrated in Fig. 2.3 Plate 1 will cost about £3 500 and a transfer press of similar capacity would cost almost as much. Consequently, there are quite a large number of injection moulding machines of this size in service. The rubber moulder is, however, often faced with the production of much heavier mouldings. If we are considering machines of 2-3 lb. capacity, the injection moulding machine is at some disadvantage. The price is in the order of £18-20 000 while an equivalent transfer press will cost about £5 000. Sometimes there is still justification for the more expensive machine owing to the shorter cycle time but only when the required volume of production is large. In most cases production requirements are such as to favour the transfer press. When even larger shots of 10-15 lb. are involved, the argument is weighted even more heavily in favour of the transfer press

If loose cores or metal inserts are to be incorporated, the horizontal injection machine, in which the mould split line is vertical, is not very convenient. In this case a combination of vertical press and screw injection unit is often used. Such a machine is illustrated in Fig. 2.5.

In this case a 60 ton press is used in conjunction with a 4–5 oz. screw injection unit, although other combinations are possible.

This machine is more expensive than the horizontal injection moulding machine, and its use is therefore restricted to applications where its special features are essential. Apart from mouldings incorporating inserts, it is useful when mouldings of 'bellows' type are being produced. Such mouldings are produced around a core pin, and the time required to remove the mouldings at the end of the cycle would keep the machine idle for an excessive time. In such cases it is usual to have duplicate sets of cores. At the end of the cycle the cores carrying the

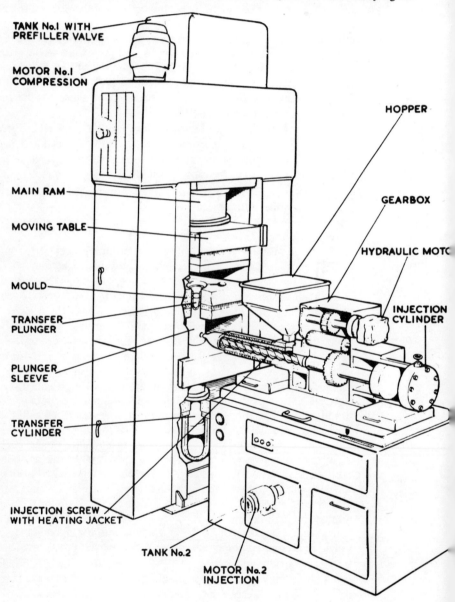

Fig. 2.5. Sectioned view of Daniels 60-ton press.

14

PLATE 1

Fig. 2.1. *(Right)*
Typical transfer press.
Courtesy of Daniels of Stroud Ltd

Fig. 2.3. *(Below)*
Single-screw injection machine
Courtesy of Alfred Herbert Ltd

PLATE 2

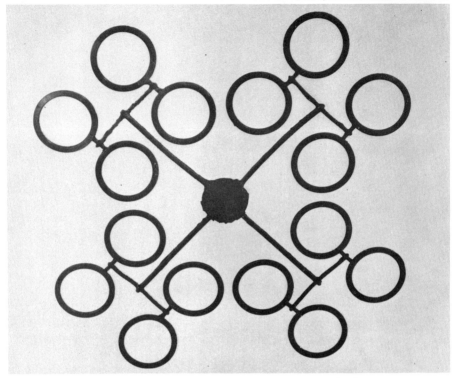

Fig. 3.5. Sixteen-cavity tensioning ring moulding having a completely balanced runner and gate system.
Photo: Edwin Peckham.

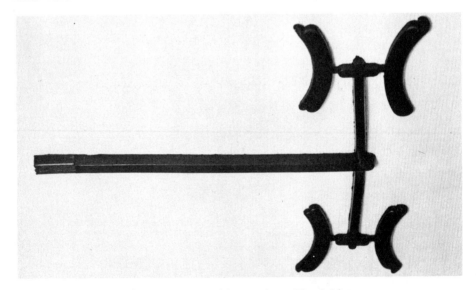

Fig. 3.6. One runner of tensioning ring (*shown above,* Fig. 3.5.) showing short mouldings. This method (making shorts) is used for checking rate of filling of all cavities for final balancing of gates. The rectangular gate used in this case was $\frac{7}{23}$ × ·060 in. depth.

PLATE 3

Fig. 3.9. Finished flexible drive spider moulding.
Photo: Edwin Peckham.

Fig. 3.10. Mouldings being ejected from the mould.

PLATE 4

Fig. 3.11. Flash-free injection moulding *vs* compression moulding.

Fig. 3.13. Six-impression car pedal mould for transfer moulding.
Daniels 'Meterjet' 100 ton press.

mouldings are removed, the duplicate set placed in the mould, and the machine immediately recycled. The cycle can be expected to take 40-60 sec. Normally this is adequate time for the removal of the mouldings, but insufficient for the cores to lose any appreciable amount of heat.

At this point some reference should be made to the quality of mouldings obtained by either of these injection processes. In the first place, the mould construction will normally be of a higher standard than would be the case with a compression mould, owing to the need to prevent flashing by the injection force. This is usually combined with a reasonable degree of mould surface finish, and the injection process will take full advantage of this in producing mouldings of equal finish. Dimensional accuracy will usually be better than in the case of compression moulding, due to the mould being locked before it is filled with rubber. Variations in size owing to differences in flash thickness are thus eliminated.

These features, although they may be better than those usually obtained from compression methods, may not always appeal to the moulder if they do not assist the moulding to perform its design function. There is one class of moulding, however, which can be expected to perform better when made by injection. This is the type of moulding incorporating a metal insert to which it is bonded. Results would appear to suggest that, probably because the rubber is in a soft, plastic condition when it meets the bonding agent, the bond strength is greater, in some cases by as much as 80%. This is particularly so when a wax antiozonant additive is used. Unless the compound has been recently mixed, the migration of this wax to the surface will reduce the bond strength. When the injection process is used the compound undergoes sufficient mixing during its passage through the runner system to re-disperse the wax, with a resultant improvement in performance. Fears that have been expressed regarding the effect of high moulding temperatures on the bonding agent, and the possibility of the adhesive being 'wiped away' by the rubber being injected under pressure seem to be unjustified.

Clearly, injection moulding must suffer from some disadvantages when compared with more traditional techniques, otherwise its use would be universal. It has to be accepted that in some cases injection moulding is more expensive; in such cases it is obviously wrong to use it. On the other hand there are many cases where the method is justified. Time does not permit a detailed discussion of costing systems, and in any case overheads, labour rates, and plant amortization rates will vary between companies. However, certain fundamentals will apply.

2.4 COST CALCULATIONS

In the cost of any moulding there will be a charge for the raw material, a machine and mould cost element, and a labour charge. Raw material will be similar in each case and can be excluded from this general review. Machinery cost will usually be relatively small, although the injection machine is likely to be most expensive. The injection mould will cost several times more per cavity than the compression mould, although, owing to the shorter cycle time, a given volume of production will be obtained with fewer cavities. Nevertheless, the injection mould will usually be more expensive. How much this affects the cost of the moulding will depend on the number of mouldings required. This means that injection moulding is favoured for longer runs. In any case, it is not likely that an expensive mould will be purchased if only a few hundred mouldings are needed. It is when we look at the labour charge that we can expect to find economies. Even here, it is most difficult to quote specific figures, and any attempt to do so would inevitably be somewhat controversial. However, let us consider a hypothetical moulding of 20 sq. in. area, made from a compound requiring about 1 600 psi moulding pressure when made by compression. If we consider the use of a 100 ton press, we could mould 6 impressions per daylight; if we assume a direct labour charge of 10s per hr, and a cycle time of 20 min., the cost of an operator looking after 4 daylights, producing 72 mouldings per hr would work out at 1·7d per moulding.

If this is compared with a transfer press of the same capacity, we would expect to have 4 cavities with a cycle time of 5 min., and, owing to the automatic nature of the machine, an operator could look after 2 machines. In this case the labour charge works out at 1·2d per moulding. On this basis for every £100 difference in mould cost we would have to produce about 50 000 mouldings before showing an advantage for the transfer press. In practice we could expect additional savings when using the transfer press in respect of blank preparation and finishing costs.

Obviously the calculations will differ for each moulding, and, while machine and mould costs play a part, the factor that is of prime importance is the number of mouldings produced per hour. Where this can be shown to be in favour of the transfer or injection machine, then regardless of machine and mould cost there will be a volume of production above which the reduced labour content will show a saving.

There are many other factors involved; for instance the fact that moulds are attached to the press and no longer have to be handled with consequent operator fatigue and mould damage; the possibility of using female labour and the changing circumstances resulting from the ever increasing cost of labour. The whole subject is clearly a complex one.

CHAPTER 3

Basic principles in the design of injection and transfer moulds

K. J. TURK, C.Eng., M.I.Mech.E.
Daniels of Stroud Ltd

3.1 INTRODUCTION

THE acceptance in the rubber industry of injection and transfer moulding of elastomers as another aid to production of rubber components is rapidly becoming commonplace, and it is necessary to have some knowledge of the basic principles in the design of moulds used for this method of manufacture. One cannot stress sufficiently the importance of mould design, as it is at this juncture of the process that the finished or almost finished components will emerge.

Injection and transfer moulds are very similar and, from a design point of view their basic requirements are the same, the only exception being the method by which the rubber is offered or fed into the mould, this being as shown diagramatically in Figs 3.1 and 3.2.

Figure 3.1 shows an injection machine screw preplasticizing unit whereby a preheated charge of rubber is injected, under pressure, into the mould via a relatively small nozzle orifice; the nozzle being in direct contact with the mould sprue feed.

Figure 3.2 shows the transfer system where a slug of rubber which has been inserted into the transfer pot prior to closure of the mould is transferred or injected under pressure directly into the mould; the slug being possibly preheated.

The approach to a mould design for any component can be varied, but it is possible, by giving careful consideration to a series of basic principles, to evolve a design which gives the best possible results, bearing in mind the final component requirements and the economics of production. Having selected a component for injection or transfer moulding it will be necessary to select the size of machine to be utilized and this will be dependent on some or all of the following:

(a) The physical dimensions and volume requirements for the component.

(b) Number of mouldings to be made per heat or shot.

(c) The projected area to be moulded versus mould locking force requirements to avoid or reduce flash to a minimum.

(d) The injection or transfer pressures necessary on the rubber whilst filling the mould cavities. This will be dependent on the flowability of the rubber compound and the nature of the moulding.

Photographic illustrations for this chapter will be found on Plates 2, 3, 4 (between pages 14 and 15) and Plates 5, 6 (between pages 70 and 71) and Plate 7 (facing page 38)

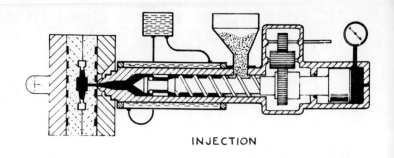

INJECTION

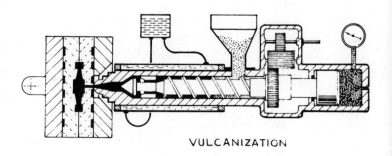

VULCANIZATION

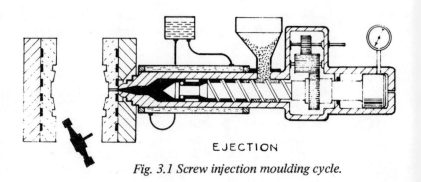

EJECTION

Fig. 3.1 Screw injection moulding cycle.

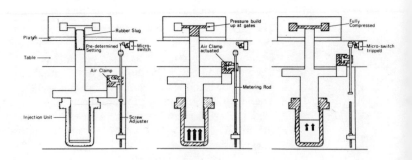

Fig. 3.2 Daniels 'Meterjet' system.

18

Once these points have been ascertained we are in a position to proceed with the mould design proper.

3.2 RUNNERS AND GATES

Runners are the channelways provided for conveying the rubber from the source of supply to the mould cavities. Various types are illustrated in Fig. 3.3 In either injection or transfer these runners radiate out from the central feed points known as a sprue for injection and a biscuit for transfer. Runners should be as short and direct as possible, depending

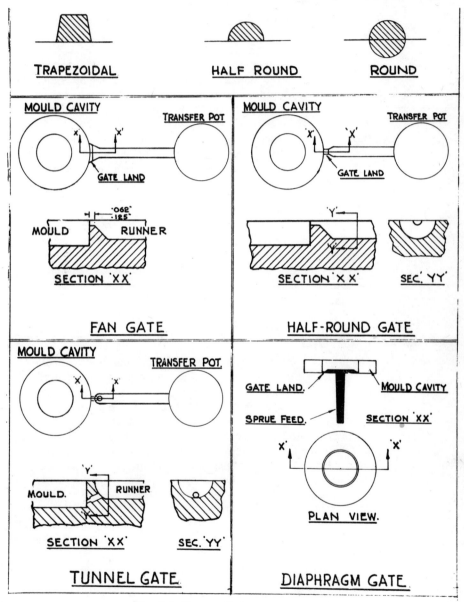

Fig. 3.3. Types of runners and gates.

on the mould cavity layout, and the best sections for obtaining good flow are full-round or trapezoidal, although half-round and rectangular sections can be used with success. It is desirable that the path taken by runners be streamlined wherever possible to promote flow and reduce pressure losses by the use of radiused corners for a change of direction. This is not always a practicable proposition as the extra mould costs involved outweigh the saving in mould efficiency. The sizes of runners are dependent on the volume and flowability of the rubber to be conveyed to the cavities. They should be kept as small as is practicable, as one must bear in mind that all material utilized in the runners plus the gates and main feed points will be wasted.

The runners terminate just short of the mould cavities and material is allowed to enter the cavity via a restrictive passageway known as the gate. There are no restrictions on the types of gate which can be employed but, naturally, the gate which is adopted should match the characteristics of the moulding for its position, for the material flow path during cavity filling and from size point of view. The runners, which are invariably larger than the gate feed points, should be diminished gradually down to the size of the gate for smooth streamline flow. The size and length of the gate lands can also be tailored to meet the article requirements and, at the same time, be such as to elevate the temperature of the rubber stock entering the cavities to a level which will reduce the cure time. Gate sections in the same form as for runners with reduced physical dimensions are generally employed and gates in the order of ·010 in. to ·015 in. thick or $\frac{1}{32}$ in. diameter have been used with success. Care must, however, be taken in the final sizing of the gate, usually carried out by practical moulding tests, to ensure that excessive work heat on the rubber does not give rise to precured rubber entering the cavity during filling. Excessive build-up of heat at the gates can also cause stress concentrations and reversion of the rubber around the feed point area of the mouldings.

3.3 CAVITY LAYOUT

Layout of the cavities for multi-impression moulds is extremely important and it is desirable that their configuration be arranged symmetrically around the central feed point so that a balanced cavity feed system can be achieved. Generally speaking, the flow of elastomers is such that the spaces or cavities are filled in order of their appearance in the flow path and it is essential that all cavities are filled at the same rate. With the two stage injection pressure systems now used and arranged to give a primary high pressure to fill the cavities and a secondary pressure for holding once the cavities are filled, it is apparent that with an unbalanced system of feeding some cavities will be pressurized with the higher injection pressure before the others are filled. This will tend to give rise to a variation in the mouldings and also possible flashing problems, particularly if the projected area is high in ratio to the locking force available.

Some typical geometrical layouts for cavities are illustrated in Fig. 3.4.

The final balancing of the mould so that all cavities fill at the same time is achieved by slight modifications to the physical dimensions of the

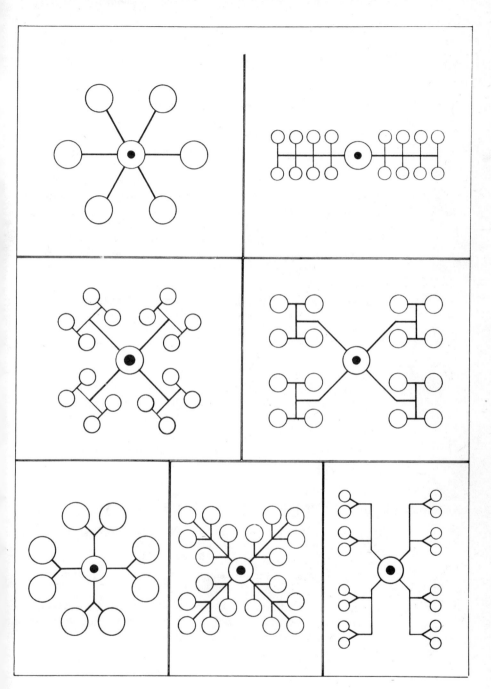

Fig. 3.4. Geometrical configuration for cavities.

21

gate feeds, the changes being decided after carrying out preliminary moulding trials and making short mouldings. Examples are given in Figs 3.5 Plate 2 and 3.6 Plate 2.

3.4 AIR VENTS AND TRIM/TEAR OFF GROOVES

It is essential to provide escape passages for the air and possible volatiles from the cavities. This is very often achieved by providing narrow shallow grooves from the cavities to atmosphere, their position invariably being sited on the split line of the mould and on the opposite side to the gate or feed points.

The size of these grooves will vary, but something in the order of 0·002 in. deep $\times$ ⅛ in. wide will suffice to leave such a small witness on the moulding that it may generally be ignored. In some cases the shape of the moulding is such that air trapping occurs away from the mould split line and, in this case, some deep buried self-clearing passageways have to be provided, or alternatively, a vacuum system attached to exhaust sufficient air when the mould closes to allow correct filling. The siting of breathers and their physical dimensions are very often provided after preliminary moulding trials have been carried out.

As is well known the problem of trying to achieve 'flash free' mouldings is always troublesome and the tediousness of after-trimming of flash, particulary if it is very thin, is undersirable. The use of trim/tear off grooves provided around the cavities can be used with success to give fairly clean mouldings and avoidance of extra after-trimming operations. The rubber is allowed to flow out through very narrow lands between the cavities and the trim grooves—the thickness of rubber across the land being so thin that on removal of the mouldings it is possible to tear off the excess rubber anchored to the mouldings and formed by the grooved cavities. Advantage can also be taken of these trim grooves for supplying additional venting points.

A typical layout of a transfer mould incorporating these features is shown in Fig. 3.7.

3.5 MOULD SPLIT LINE CLAMPING FACES

To minimize flash it is essential to achieve a good bite between the split line faces of the mould when the clamping force is applied. This can be achieved by providing land faces around the feed, runner, gate and cavity areas, the surface pressures on the total land area being at a safe stress to avoid excessive hobbing.

The importance of flatness and surface finish on the mould split line faces cannot be over emphasized and the use of super ground or lapped faces is essential to achieve the best results.

3.6 MOULD HEATING

This is normally achieved by the use of electricity and temperature is controlled in the region of 140–200°C by the use of standard proportioning thermostatic controllers.

In the case of moulds fitted into vertical clamp force machines the heat

PLATE 5

Fig. 3.14. Short gasket mouldings.
Photo: Edwin Peckham.

Fig. 3.15. Complete gasket mouldings.
Photo: Edwin Peckham.

PLATE 6

Fig. 3.16. Split line feed.

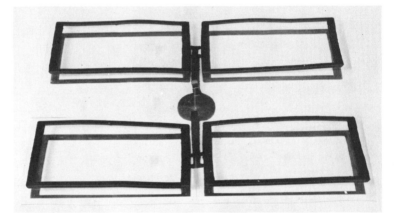

Fig. 3.17. Two-deck system moulding.
Photo: Edwin Peckham.

Fig. 3.18. Sixteen-impression bush moulding.
Courtesy: Dunlop Co. Ltd.

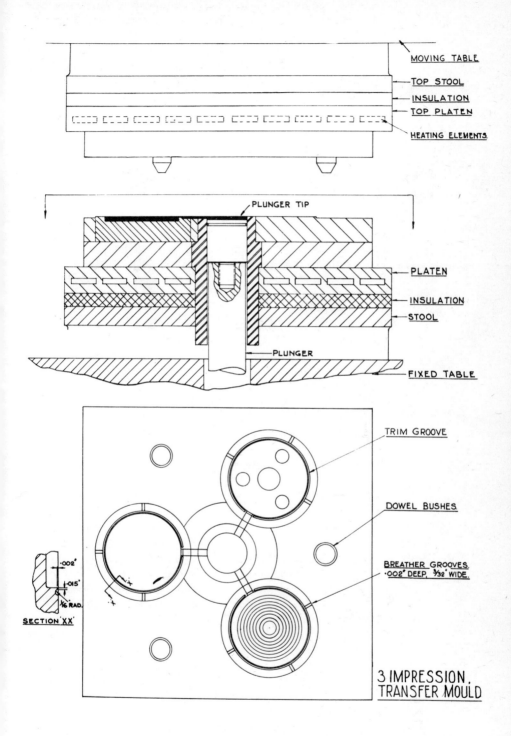

MOVING TABLE
TOP STOOL
INSULATION
TOP PLATEN
HEATING ELEMENTS.

PLUNGER TIP

PLATEN

INSULATION

STOOL

PLUNGER

FIXED TABLE

TRIM GROOVE

DOWEL BUSHES.

BREATHER GROOVES.
·002" DEEP, ³⁄₃₂" WIDE.

·002"
·015"
¹⁄₆ RAD.
SECTION 'XX'

3 IMPRESSION,
TRANSFER MOULD

Fig. 3.7. Layout of a transfer mould producing three gaskets of different configurations.

23

is usually supplied from platens which are supplied with the machine and to which the moulds are bolted, as for compression moulding techniques.

Moulds fitted to the horizontal in-line injection machines are usually designed with built-in heaters, the type of heaters used being dependent on their design. The most commonly used elements are strip, cartridge or band type if the outside of the mould is circular.

When considering mould designs from a heating point of view an even heat distribution is of paramount importance and the positioning of heating elements too close to cavities should be avoided as this will give rise to undesirable hot spots.

3.7 EJECTION

Although most rubber mouldings do not lend themselves readily to automatic ejection, it is sometimes desirable to provide means for assistance in removal of the mouldings, by providing ejector mechanisms to retain the moulding in the correct half of the mould for removal. Most injection machinery is, or can be, provided with mechanical or hydraulic mechanisms which can be suitably connected to the mould for this purpose.

Ejection is generally more suited to thick section mouldings with rubbers having good hot strength and mouldings with bonded metal inserts or core pins whereby ejection pins may be suitably arranged to push on the metal inserts. When ejection pins are used to push directly on the rubber, the use of mushroom or mitre type pins is recommended, as, with the use of straight pins, the rubber tends to flow down them via even the smallest clearances to cause eventual stiction or seizure.

3.8 GENERAL MOULD DESIGN REQUIREMENTS AND COMMENTS

1 Moulds should be as rigid as is practicable to withstand first, the injection pressures, which can be as high as 20 000 lb. psi and secondly, the avoidance of deflections which eventually give rise to flash.

2 If guide dowels and bushes are utilized for line up of the mould halves the dowels should be provided with a liberal taper for entry into the bushes and be of sufficient rigidity to perform their function. Care should also be taken in sizing them to allow for thermal expansion.

3 The working faces of the mould, particularly on the split line or closure faces, should be kept as clear as possible of screw fixing holes and joints, as these become sources for a build-up of rubber causing flashing, and traps for foreign matter; this will eventually cause mould damage.

4 For long life in production the use of good quality toughened steels is advisable, particularly for the areas which come into contact with the rubber.

5 To take full advantage of the high quality of moulding by injection

or transfer techniques it is desirable that all surfaces coming into contact with the rubber be highly polished and the cavities and core pins preferably plated.

6   Shrinkage allowances must, of course, be made in the cavities and on the core pins and these will be dependent on the type of rubber compound being used plus all the variables in processing including mould surface temperatures, injection pressures and cure times. Here again, allowances should be made in the mould manufacture so that final sizing of the cavities and core pins, can if necessary, be carried out after initial testing and proving. To assist in sizing evaluation it is desirable in the case of multi-cavity moulds to have identification numbers stamped in the cavities or on the core pins.

7   In the design of moulds including those of the most comprehensive type the means for handling the mould into and out of the machine is more often than not forgotten. It is usually quite a simple matter to provide some additional tappings or holes so that suitable attachments for lifting facilities can be fitted, which can avoid untold damage to both mould and machine.

The general principles for mould design have now been outlined. Further information may be given with the aid of a number of illustrations.

Figure 3.8 shows a two-impression mould for the manufacture of flexible drive spiders using a horizontal in-line screw injection moulding machine. Automatic ejection was used and achieved by means of an ejector grid and an ejector pin on each tooth of the moulding. A calico and hemp fibre filled nitrile compound was employed. Figure 3.9 Plate 3, shows some completed mouldings and their assembly in the metal drive dog. Figure 3.10 Plate 3, shows an actual mould with the mouldings being ejected using air assistance and free fall. A 2½ in. overall cycle was employed, no after-trimming being required except

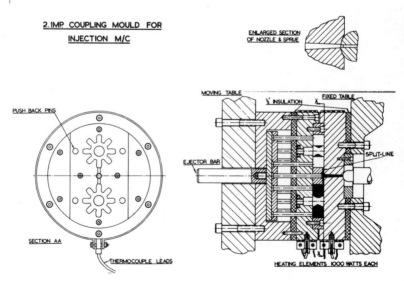

Fig. 3.8. Mould for flexible drive spiders.

that of removing the gate feed. This point is further illustrated in
Fig. 3.11 Plate 4, where a comparison is made between a flash-free
injection moulded piece and a compression moulded piece.

Figure 3.12 shows a six-impression transfer mould for circular
disc mouldings with centre steel insert. A bottom ejection system is
incorporated. A similar transfer mould for a car pedal is shown in
Fig. 3.13 Plate 4.

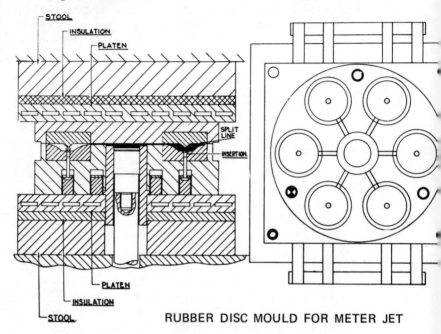

RUBBER DISC MOULD FOR METER JET

Fig. 3.12. Six-impression transfer mould.

Figure 3.14 Plate 5, indicates how rubber is filling the mould by making
short mouldings of various gaskets. Figure 3.15 Plate 5, shows the
complete mouldings in which the trim and breathe grooves should be
noted. Figure 3.16 Plate 6, illustrates joint gaskets with a split line feed
and consequent difficulties.

To conclude, a few injection and transfer moulded articles may be
illustrated. Figure 3.17 Plate 6, shows an eight-impression rectangular
gasket moulding, three plate mould, 'Piggy-Back' or two-deck system.
Figure 3.18 Plate 6, illustrates a sixteen-impression bush moulding by
the transfer method using automatic core pulling. Figure 3.19 Plate 7,
shows a car gear shift lever bellows and cable escutcheon while
Fig. 3.20 Plate 7, shows pipe joint rings where trim grooves
were employed.

These various mouldings show the versatility of the injection moulding
processes and that little difficulty should be experienced in
manufacturing the most complex items.

ACKNOWLEDGEMENTS

I would like to acknowledge grateful thanks to T. H. & J. Daniels Ltd
of Stroud, Gloucestershire, for making available the basic information
from which this paper was prepared.

DISCUSSION

P. BOIS *(Industrial Plastics Ltd)*: Can weld lines in open (i.e. ring type mouldings) be eliminated by air bleeding–what percentage weld line strengths can be expected?

ANSWER: The use of air bleeding or venting points for cavities is essential to reduce or eliminate visual weld line effects and improve weld line strengths. If no provision is allowed for air escape, the air in any cavity will be compressed into the last spaces to fill and this will create temperature rises which will be sufficient to cure the surfaces of the rubber which are joining together; hence poor and weak weld lines.

Weld line strengths in order of 90–95% can be satisfactorily achieved provided the optimum processing conditions for mould surface temperature, runner and gate sizes and speeds of injection are balanced with the rubber compound curing system. The ideal condition is to arrange for the cavities to be completely filled before the final stages of cure of the rubber compound are allowed to take place.

P. H. KELLETT *(SATRA)*: In the footwear industry where pvc soles are injected directly on to shoe uppers, high locking pressures are avoided by preventing build-up of injection pressure in the mould by means of a pressure sensitive switch in the mould surface. This switch stops injection and so pressure developed inside the mould can never exceed the locking pressure. With this technique very low locking pressures can be used.

Why is this technique not used in rubber injection machines?

A: In certain applications it may be possible to use a pressure sensitive switch built into the mould cavity, provided it will withstand constant mould temperatures of up to 200°C and that it can be so sited that rubber does not find its way into the actuating mechanism to give rise to failure of operation and an added mould complication.

It must be stressed that a mould for pvc (thermoplastic) is relatively cool and that the material injected is freezing off, whereas with rubbers they can be quite fluidized for a while in the mould, before cure takes place, and this, together with the thermal expansions of rubber generally speaking demands higher locking pressures.

DR L. J. GERHARDT *(Vitamol Precision)*: Why is injection pressure reduced after the mould has been filled and not just before it is filled?

A: Injection pressures are reduced immediately the cavities are filled to avoid stress concentrations around the feed point areas of the mouldings (gates). If the high injection pressures necessary for flowing the material through the runners and gates, and sometimes cavities, are maintained, there is the tendency for more material to be forced into the cavities whilst cure is taking place and this can give rise to lamination effects.

It can be arranged for the reduced injection pressure to be applied before the cavities are filled provided the reduced pressure is sufficient to complete cavity filling and avoid the aforementioned stress concentration effects.

QUESTION: Are half-round runners less satisfactory than trapezoid or round runners?

A: Trapezoidal and round runners offer the least resistance to material flow but half-round runners can be used with success, particularly if their cross-sectional area is equivalent to a full round runner.

Q: Is the sprue balanced against runners and are tapered sprues and runners used?

A: The sprue feed of a screw/ram injection machine is not always balanced in cross-sectional area with respect to the runners used. More often the sprue feed, from the machine and mould point of view, is less and, owing to its relatively short length it can be used to advantage by putting extra work heat into the rubber.

Tapered sprues are advisable to assist in the release of the sprue from the nozzle when the mould is opened for ejection of the mouldings. At the opposite end to the nozzle of the sprue feed a dovetailed type cavity is also provided in the moving half of the mould to act as a puller for the sprue from the nozzle orifice.

Q: Why are some of the runners shown so large?

A: Large runners are sometimes used to promote flow and reduce pressure losses through them. The sizes of the runners are very dependent on the flowability and nature of the compound being used and preliminary tests with a compound are often carried out before runner sizes are fixed.

Q: Are not multiple feeds for the same cavity trapping the air?

A: The use of multiple gate feeds for the same cavity does not give rise to air-trapping problems provided suitable breather points to atmosphere are supplied. With multi-gate feeds the equivalent number of weld lines occurs and it is necessary to provide breathe grooves at these weld line points.

W. S. PENN *(Borough Polytechnic)*: Which factors lead to low flash in injection moulding?

A: Mouldings shown in seven of the figures were made using the trim/tear off groove system, and the rubber formed by these grooves has been removed to show the quality of the finished components; this being easily removed by hand tearing from the moulding by the machine operator between cycles. The other mouldings shown were virtually flash free and these were obtained by the use of good quality moulds and correct adjustment and setting of the machines.

Factors leading to *virtually* 'flash-free' mouldings can be attributed to some or all of the following:

1 The use of first-class moulds made in toughened steels and of sufficient robustness to withstand the injection pressures exerted upon them.

2 Mould materials should be stress relieved before final sizing to avoid heat distortion and the clamping or split line faces should be super ground or lapped.

3 Provision of land clamping faces around the mould cavities, runners, gates and sprue or transfer areas to ensure a good bite on mould closure faces.

4 In the case of multi-cavity moulds ensure that cavity filling balance is achieved so that the primary high injection pressure

for filling and the secondary reduced injection hold-on pressure once the cavities are filled can be correctly adjusted to avoid the mould locking force being overcome.

5   Correct utilization of all machine adjustments, including mould surface temperature, injection pressures or speed of injection, volume charge control, and the preheat temperature of rubber stock.

6   Avoidance of damage to mould split line or closure faces and periodic cleaning of faces.

J. R. WHISTON *(Dunlop Co. Ltd–Polymer Engineering Division):* The economics quoted by Mr Gardner are questionable because they do not take into account that the whole of the platen area may be utilized. Injection moulding is restricted by the projected area usable which, if calculated by equating the locking force available to the product of projected area and injection pressure, is very small. The theoretical projected area is much less than actually achievable. Is there an established factor which may be applied to the theoretical value which gives an estimate of the actual achievable?

A: Any typical examples of economics that are quoted can be a topic for extensive discussion. However, there are many variables which control the amount of usable platen area and even with compression moulding techniques the best results are very often achieved by limiting the amount of area utilized.

The main parameters controlling the actual projected moulded areas achievable by injection or transfer moulding systems are:

1   Nature of product (thick or thin sections and intricacy of shape)

2   Nature of rubber compound (flowability, type of fillers and cure system).

3   Injection pressures. As already mentioned, most modern injection machines have two-stage injection pressure systems whereby a primary high injection pressure can be utilized for filling the cavities and a secondary reduced pressure for holding, which is initiated immediately the cavities are filled; both pressures being adjustable. I would also mention that the theoretical injection pressure indicated during the filling of cavities is only applicable to the feed, runner and gate areas and that the pressure in the cavities is invariably much lower.

To substantiate these comments I can give an example where mouldings having a total projected area of 62 sq. in. are being satisfactorily moulded on a 45 ton clamp force machine; the theoretical indicated pressure on the rubber compound during filling being 7·4 tons psi, whilst the reduced holding pressure was 0·7 tons psi.

I do not know of any established factor which can be applied to the theoretical projected area to give the actual possible. However, from personal experience with many types of rubber compounds and products I have found that a projected area in sq. in. equal to half the mould locking force in tons acts as a very good guide, i.e. a projected area of 100 sq. in. achievable in a 200 ton lock force machine.

CHAPTER 4

Basic machine layout and its effect on output

P. F. HARRISON, C.Eng., M.I.Prod.E.
GKN Machinery Ltd–Peco Division

4.1 INTRODUCTION

IN order to use injection moulding competitively in the rubber industry, the increased performance of the process must successfully offset its increased capital expenditure compared with that of earlier, simpler processes. The most obvious way in which the expensive injection moulding machine can pay for itself is by an increased rate of output resulting from reduced cure times. The selection of the correct machine layout for the duty intended, however, can optimize the utilization of each part of the machine and therefore, each unit of capital expenditure. Let us consider different basic machine layouts and their effective utilization.

4.2 SINGLE LOCKING UNIT/SINGLE INJECTION UNIT

In this, the most simple form of conventional injection moulding machine, the injection time and the cure time form the working cycle. Here useful processing is accomplished but the total machine cycle is made up from the injection time, the cure time, plus any extension of the cure time necessary for the completion of preplasticizing with the addition of mould opening, ejection, mould cleaning and insert loading, followed by mould closing.

It will be seen in Fig. 4.1 that this type of machine, because of its simplicity has the lowest capital expenditure of any 'true' injection moulding process. Providing the cure time is reasonably short the injection unit is usefully employed during the cure time in precharging or preplasticizing. The time which elapses from the completion of cure until the commencement of the next injection is wasted time during which no useful processing takes place. In the simple design in question, this wasted time will be extended by slow platen movement, insert loading, difficulties in stripping or ejection and any hand work necessary on the mould. The simple, conventional injection moulding machine therefore has the advantage of low cost and is an efficient solution to moulding problems when cure time and precharge time are reasonably similar and mould hand work is at a minimum.

4.3 ONE INJECTION UNIT FEEDING SEVERAL LOCKING UNIT PRESSES (Fig. 4.2)

In this type of plant where two or three vertical presses are fed by a single mobile injection unit the first cost is considerably greater than for the simple machine discussed in Section 4.2. The increase in cost is more than a *pro rata* increase of locking unit value owing to the increased control gear complexity, necessary to phase the operation of the injection unit. Each locking unit is a heavy press capable of developing the full locking force specified and in most practical examples of this construction the limitation of split line injection is accepted. The object of this design is a more efficient utilization of the injection unit and it is, therefore, most likely to succeed where cure times are long compared with the cycle of operation of the injection assembly or where extended hand working or insert loading is demanded by the mould. The injection unit will be fully occupied when its full cycle time is equal to a typical cure, mould open/close and mould hand work time less the injection traverse time all divided by the number of locking units.

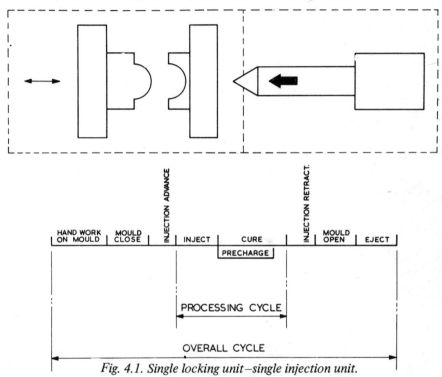

Fig. 4.1. Single locking unit—single injection unit.

4.4 ROTARY TABLE MACHINE WITH FULL LOCKING FORCE AT EACH STATION AND ONE OR MORE INJECTION UNITS

Machines of this type are highly expensive and complex pieces of plant suitable for the production of items with relatively long curing times; sometimes two dissimilar compounds are injected by means of separate injection units. As employed in the rubber boot industry a fairly extensive hand extraction period is needed and the rotary table principle ensures full utilization of the injection unit, adequate curing time with moulds locked and unloading positions for removing the complex mould components.

Figure 4.3 illustrates a rotary table machine in diagrammatic form and in order to simplify the timing diagram only four mould stations are

31

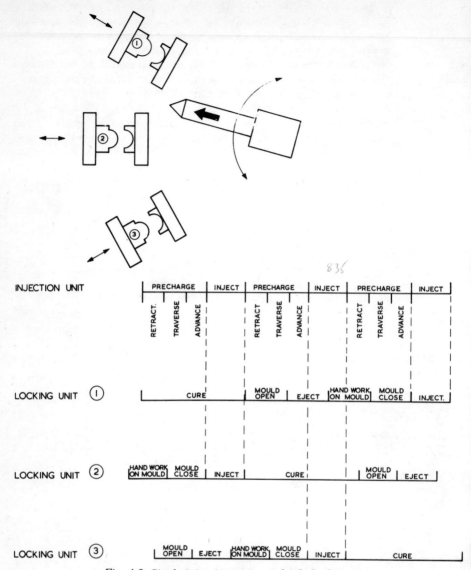

835

INJECTION UNIT		PRECHARGE			INJECT	PRECHARGE			INJECT	PRECHARGE			INJECT
		RETRACT.	TRAVERSE	ADVANCE		RETRACT	TRAVERSE	ADVANCE		RETRACT	TRAVERSE	ADVANCE	
LOCKING UNIT ①		CURE				MOULD OPEN		EJECT	HAND WORK ON MOULD	MOULD CLOSE		INJECT.	
LOCKING UNIT ②	HAND WORK ON MOULD	MOULD CLOSE	INJECT		CURE				MOULD OPEN	EJECT			
LOCKING UNIT ③	MOULD OPEN	EJECT	HAND WORK ON MOULD	MOULD CLOSE	INJECT	CURE							

Fig. 4.2. Single injection unit–multiple locking unit.

shown. It can be seen that the full occupation of the injection unit will be achieved at the same point as that in the layout shown in Fig. 4.2.

In heavy duty machines of this type with provision for multiple injection units and complex mould arrangements, each station of the rotary table carries a locking unit which can exert the full specified locking force for the whole period over which the mould faces are together and therefore, during the whole cure time. Simpler machines adapted for use with one injection unit only and, where the application of locking force is not necessary for the whole of the cure period, are built with the locking arrangements opposite the injection unit index position only. Light clamps or the effects of gravity keep the moulds together during most of the cure time.

32

It can be seen that with this type of design maximum utilization of all the expensive machine components, i.e. injection unit and heavy locking parts, is achieved at the expense of mechanical and control complexity in the lighter operations of table indexing, mould opening and closing, etc. In some cases the mould opening and closing arrangements are by hand.

4.5. MOULD OPENED REMOTELY FROM MACHINE

The fourth arrangement (Fig. 4.4) for consideration is that in which the mould opening, ejecting or stripping and any hand work are carried out

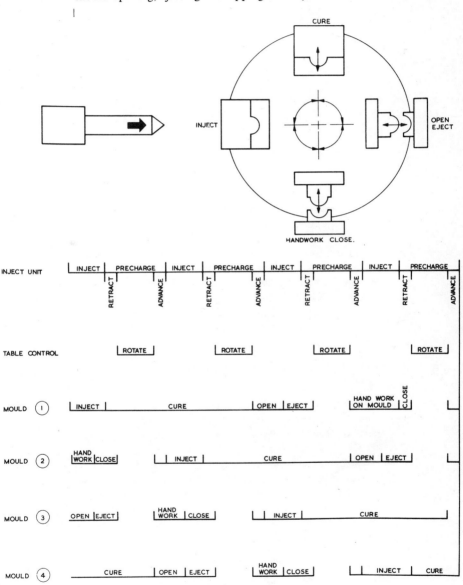

Fig. 4.3. Rotary mould table—single injection unit.

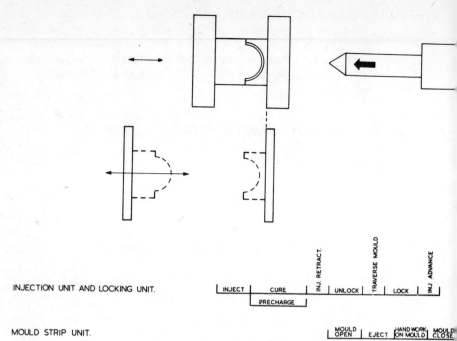

INJECTION UNIT AND LOCKING UNIT.

INJECT	CURE	INJ. RETRACT.	UNLOCK	TRAVERSE MOULD	LOCK	INJ ADVANCE
	PRECHARGE					

MOULD STRIP UNIT.

| MOULD OPEN | EJECT | HAND WORK ON MOULD | MOULD CLOSE |

Fig. 4.4. Separate mould strip unit.

with the mould removed from the press. This occurs particularly in the case of existing vertical presses adapted for use with an injection moulding unit. Quite often the moulds are opened and stripped by hand and the overall economics of the process depend on the number of moulds in use, their relative size and the proportions of cure time to stripping and hand work time. The system is of interest, however, in demonstrating that the opening and closing function may be separated from the heavy locking unit parts and this is a feature of the next construction to be considered.

4.6. SINGLE LOCKING AND INJECTION STATION WITH REMOTE STRIPPING UNITS

In the construction shown in Fig. 4.5, a single central locking unit and injection assembly is flanked by two mould opening and stripping units making three 'press' assemblies in line. Two moulds mounted on a moving carriage alternate between the common, central injection and locking unit and their individual stripping units on the appropriate side. It can be seen that this construction gives an advantage when the mould opening and mould hand work time is long compared with the total injection and cure time. In practice, a single injection and locking unit with two stripping units can give the output of two simple machines. In addition, as the long mould opening stroke is divorced from the heavy locking force an economic construction results. As in the case of rotary table and multiple locking machines, however, complete automation of this arrangement implies a considerable increase in control and power unit complexity, compared with a simple machine.

34

4.7. REMOTE CURING IN A SEPARATE PRESS

The final arrangement to be considered is similar to that shown in Fig. 4.5. It will be appreciated that in this arrangement where the moulds are removed from the machine for opening and stripping they may be removed while still curing, providing their shape and section does not demand full locking force during this phase. If locking force is required during curing, however, a possible arrangement is to transfer the moulds from the injection moulding machine to separate heated plateu simple compression type presses. In this way complete utilization of the injection unit can be achieved.

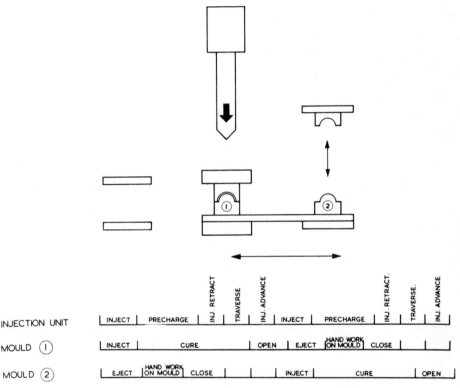

Fig. 4.5. *Central locking unit and side stripping stations.*

Machine cycles illustrated where multi-locking unit systems are shown are arranged as used with a shut off nozzle for optimum speed.

It can be seen that many interlocking features influence the choice of the injection moulding machine layout for rubber work. Screw preplasticizers give the benefit of short cure times but themselves demand a precharge time. This must be accommodated in the injection unit cycle. Plunger injection units on the other hand, give lower benefits in terms of reduced cure time but need no precharging except perhaps a very short delay while a slug is loaded.

Machines in which the moulds are moved from platen to platen are, for practical purposes, limited to heating by contact with hot platens or a the very best to heating by induction processes. On the other hand,

35

heating by hot oil or electrical elements within the tool, with the advantages of more accurate control where it is most needed, are available in machines of the first three categories described, where the mould and platen are permanently connected during the cycle. Finally, it is instructive to consider the effects on the cycle times and, consequently, on the optimum machine layout of advances in compounding which will reduce cure and possible injection times and also the effect of increased automation in mould ejection, stripping and the loading of inserts.

In this short chapter the author has attempted to show the various mechanical solutions to the problem of economically providing the machine features required for rubber moulding. In the provision of these facilities with the best compromise of cost, performance, and reliability the machine design engineer can make his contribution to the development of the process.

DISCUSSION

W. S. PENN *(Borough Polytechnic)*: Could you please indicate the relative usage of electrical/air and fluid heating/cooling used in injection machines for elastomers. What are their merits and demerits?

ANSWER: Most injection moulding machines for rubber which employ a single-screw injection unit use a form of barrel temperature control which includes a fluid circulating heat transfer medium. Only a small minority of these machines use electric resistance heaters; this is in contrast to thermoplastic practice. The fluid heat transfer system has a high thermal inertia and gives good stability at low temperatures and provides heating and cooling with a single control. It is, therefore, most suitable for use in rubber processing. For thermoplastic processing, electrical resistance heaters provide greater heater power, higher temperature working and a greater facility for establishing a temperature profile along the barrel.

P. BOIS *(Industrial Plastics Ltd)*: Is the screw ram type of machine really the only practical configuration? What of the benefits of transferring the material in a single shot quantity to a separate transfer cylinder?

A: The screw ram plasticizing unit offers the outstanding advantage of a simple flow path and the maximum utilization of the self cleaning and mixing action of the screw. It has proved to be the most practical layout for rubber, thermoset and heat sensitive thermoplastic processing. The screw and transfer pot system offers the advantage of a continuously running nonreciprocating screw, whose design can be established irrespective of shot displacement conditions but it has a disadvantage of a more complex flow path including a valve element.

R. F. POWELL *(Trist Mouldings and Seals)*: Can available injection pressure be raised with existing shot displacement another 7 000 psi and what would be the additional cost?

A: Injection pressure depends on the injection force and the cross-sectional area of the screw when used as a ram. To increase injection pressure by a significant amount, it is likely that a new cylinder unit with its associated stressed parts will be needed. It is anticipated that a replacement injection unit would cost about one-third of the cost of a complete single-station injection moulding machine.

R. F. POWELL: Would increased injection pressures for a given shot weight increase machine versatility over a wider range of compounds?

A: Increased injection pressure up to 20 000 psi will allow a wide range of compounds to be processed and will also give greater flexibility of conditions when using existing compounds. Care should be taken that other features such as screw drive power or heat transfer area do not then become a limitation.

DR L. J. GERHARDT *(Vitamol Precision)*: I cannot see any practical solution where mould is transferred after injection into multi-daylight compression presses.

A: A process is in operation in which moulds are transferred from an injection station to a curing press. It is possible that the relaxation in locking force during the transfer period offers an advantage in processing on certain components.

J. R. WHISTON *(Dunlop Co. Ltd—Polymer Engineering Division)*: Much attention has been given to the method of heating the barrel. Do you consider attention should be given to heating/cooling the screw? What advantages could be expected?

A: Screw feed provides a circulating effect within each flight which ensures that all sections of the rubber to be processed pass close to the heated barrel. In theory, a heated screw will provide a greater transfer area but the difficulties of providing controlled temperature heating in the screw, limit this advantage in practice. It has been found however, that a low power cooling effect usually achieved by air circulation assists in stabilizing the temperature of the screw and prevents curing on the screw tip or forward section which otherwise can take place several hours after the commencement of a satisfactory moulding run. Modern injection moulding machines for rubber include screw cooling as a standard feature.

CHAPTER **5**

Injection machine controls and their use

P. F. HARRISON, C.Eng., M.I.Prod.E.,
GKN Machinery Ltd–Peco Division

5.1   INTRODUCTION

IN the moulding of rubber by the established compression press
methods the process is controlled largely by changes to the compound
and to the timing and temperatures maintained during the moulding
operation. When injection moulding is employed the control of injection
pressure in several stages is added to the established process variables
while the controls of temperature and time are made more sensitive
and more effective.

The controls of the injection moulding machine may be divided into
four broad classifications:

   1 Movement selectors
   2 Mechanical adjustments
   3 Cycle pattern controls
   4 Processing variables.

These are considered in subsequent sections.

5.2   MOVEMENT SELECTORS

The first classification covers the movement controls by which machine
operations may be selected manually and which are used under normal
circumstances only during the setting or perhaps in starting the machine
production run. These controls comprise push buttons which operate
hydraulic solenoid valves or hydraulic pilot valves directly actuated by
levers. Controls for the following functions are available on a simple
single station machine:

   Mould close
   Mould open
   Inject
   Precharge or screw rotate
   Injection unit advance
   Injection unit retract.

To these controls may be added the main motor start and stop buttons
and directional controls for auxiliary movements such as core cylinders
or ejectors. In some machines the injection carriage may be selected to
operate in phase with the mould open and close as a first stage of
semi-automatic operation. In almost all designs, it is possible to progress

*The photograph used in illustration of this chapter, Fig. 5.3 will be found on
Plate 8 (facing page 39)*

38

PLATE 7

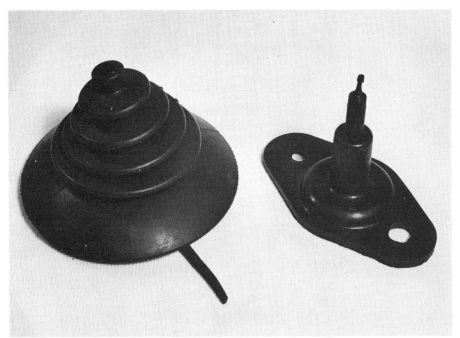

Fig. 3.19. Shift bellows and cable escutcheon.

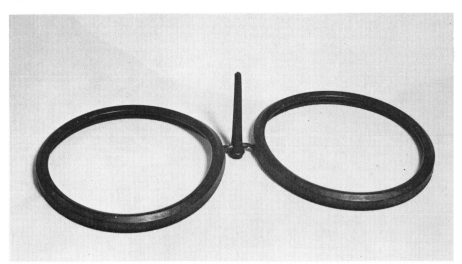

Fig. 3.20. Pipe joint rings.

PLATE 8

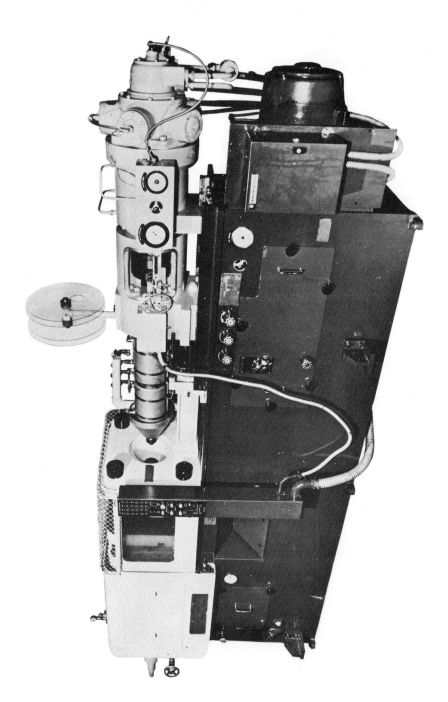

Fig. 5.3. Single station ⌐ ⌐ ⌐ moulding machine designed specifically for moulding rubber.

to the stage of automatic cycle operation, sometimes called semi-automatic operation, in which one complete machine cycle takes place after the mould close button has been selected. Many machines are adapted to fully automatic or continuous cycling and for this type of operation, some machines are fitted with a safety device which provides a low pressure to close the mould until a point is reached when the mould faces have come together cleanly without being obstructed by foreign matter. Only then is the full locking pressure applied.

5.3 MECHANICAL ADJUSTMENTS

The second category of injection moulding machine controls covers the mechanical adjustments to the machine to accommodate a variety of moulds. The settings in this category include the regulation of mould opening stroke and on toggle machines the setting of the mould thickness. Associated with these settings are the speed controls for mould movement which give cushioning at the end of the strokes. Ejectors are normally arranged so that the stroke of the ejector bar may be selected. Certain injection moulding machines have setting controls to ensure that the nozzle contacts the mould effectively. In other designs this setting is achieved automatically. As moulds of different capacity are set up for operation, it is necessary to adjust the injection volume and this adjustment is performed by a limit switch arranged to terminate the precharge movement.

5.4 CYCLE PATTERN CONTROLS

The author has placed the cycle pattern controls in the third division. This form of control is chiefly exercised as a modification of the basic machine cycle or its pattern of movement. As an example, the injection nozzle may be retracted at the beginning of the cure time or at the end of the cure time. The precharge period may be contained within the cure time or may overlap into the mould movement phase; or alternatively, its commencement may be delayed so that it is effective only during the last part of the cure time. The selection of a suitable cycle pattern while such control refinement is available often has a decided bearing on the operating conditions.

5.5 PROCESSING VARIABLES

The author has placed in the fourth category those controls which are of the greatest interest in the present context. These are the process controls which directly affect the moulding conditions.

It is obvious that in any process involving heat transfer, time must be rigidly controlled and most injection moulding machines control their process times stage by stage by several timers. A typical machine will have the following timers:

> Injection
> Injection boost
> Injection hold

Screw delay
Cure
Recycle

The injection moulding process involves a pressure filling of the mould and the pressure control during the injection phase is very critical. Control is normally arranged in two consecutive stages and this may be extended to three stages in some equipment. In rubber processing, the pressure affects the speed of mould filling and this in turn has an influence on compound temperature so that pressure has an influence on the third main process control, that of temperature.

Temperature control in the injection moulding process is achieved by a combination of settings of barrel jacket water temperature, screw back pressure, screw speed, rate of injection, nozzle diameter, tool temperature and the secondary effects of injection pressure and time settings.

All the variables in Section 5.5 may be divided into two classes.

In the first class, there are those controls which act to influence the heating and processing of the rubber during the precharging stage when the rubber is being accumulated for injection. The controls effective during this period are the barrel temperature, screw speed and screw back pressure. The second set of controls are those which act to influence the second stage of injection unit operation, that of mould filling. Settings influential in this phase are the injection pressure in its several stages, the rate of injection, the mould temperature, the timing of the pressure stages and the nozzle orifice.

5.5.1 INJECTION STAGE CONTROLS

Some variables in Section 5.5–the screw speed, back pressure injection pressure and rate of injection–are controlled hydraulically and Fig. 5.1 shows a typical hydraulic circuit for this purpose.

Oil is delivered from a pump which is normally of a fixed displacement type. Machines in which more than 10 hp are used normally have a double pump so that input power may be reserved when the machine is resting, or holding its position. Oil is delivered from the pumping section to a directional valve, shown in the centre of the illustration, which is of a two-stage type. The solenoids shown actuate a small pilot spool which directs control oil to the appropriate end of the main spool. The valves are illustrated to a convention in which the flow connections in the centred and offset positions are shown in adjoining boxes; in the case of centring four-way valve, four ports and three connection possibilities are shown. In order to understand the convention, imagine that the boxes move under stationary ports when pushed by the pilot oil. The directional four-way valve in the circuit when centred allows the injection unit to rest. The offset positions direct oil to power the injection and screw rotation functions. Pipes are shown diagrammatically as lines leading from the directional valve to the screw drive motor and the injection cylinder. In the screw drive motor line a flow control valve is indicated. This is a manually adjusted orifice supplemented by a temperature- and pressure-compensated hydrostat which provides flow regulation to control screw speed. In many modern machines, coarse speeds of screw speed regulation are established by the selection of the hydraulic drive

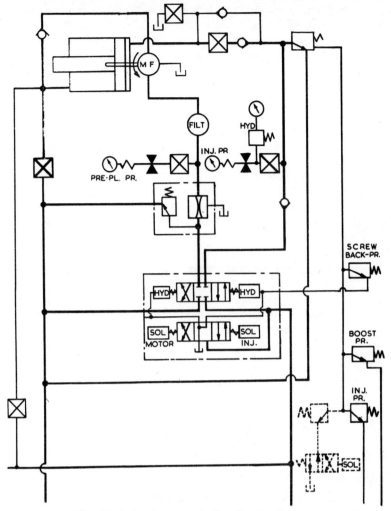

Fig. 5.1. Injection press hydraulic circuit.

pumps in addition to the fine adjustment provided by the flow control valve.

In the line leading to the injection cylinder, a manually operated restrictor without pressure compensation regulates flow. This is closely adjacent to the cylinder with a by-pass allowing free return flow.

In a branch from the injection feed line is the injection relief valve. It is the duty of this valve to regulate the injection pressure in several stages, and the screw back pressure. The valve is normally of a servo type such as the Vickers Hydrocone design. In this valve a main spool vents oil to the reservoir at a preset pressure. The pressure is set by a small servo valve mounted integrally. It is possible, however, to supplement the control of the servo valve by external units which may be piped to a convenient position of access on the machine.

The servo valve acts to relieve pressure when its setting is reached and, consequently, the servo which is set lowest out of a group controlling a single main valve will take over the control. The pilot arrangement shown is brought into action during the first two phases of the injection movement to give two levels of injection pressure.

The uppermost servo valve is piped to the pilot line which selects injection and is, therefore, held out of action during injection. This control becomes effective only when the injection pilot line is connected to the reservoir and this occurs when the screw rotate function is selected. The purpose of this valve is to regulate the back pressure, this being the pressure which opposes the 'rearward' retraction of the screw during its rotating phase. This pressure, which is normally less than a quarter of the value of the injection pressure, is a powerful control which is essential to establish the level of shear working given to the compound.

It can be seen that the screw speed and injection pressures are controlled directly and steplessly in this typical arrangement. The actual value of injection speed achieved is dependent on injection pressure and nozzle diameter.

Barrel temperature is normally controlled in rubber injection moulding machines by the use of water or oil jackets surrounding the barrel through which a thermostatically controlled heating fluid is circulated. An external unit controls the fluid temperature and use is frequently made of proprietory plastic mould temperature control units. This form of temperature control gives high inherent stability owing to the thermal inertia of the system but does not allow for a quick response to meet changing conditions nor is it easy to establish a pronounced temperature gradient along the barrel. Thermoplastic practice normally employs band heaters with on/off or proportioning control from shallow thermo-couples. This gives a much quicker response. This system has found little favour for rubber machines, however, and it is generally concluded that the fluid circulation method is preferable for the relatively low temperatures involved.

5.5.2 MOULD STAGE CONTROLS

The mould heating for injection moulding is either by the direct application of electric resistance heaters to the mould either internally in cartridge form or externally in the form of bands (Fig. 5.2). Occasionally, the moulds may be heated by hot oil circulation. In the case of some moulds, particularly those with a feed or runner system which is kept below curing temperature, two systems of temperature control may be used. In some cases, particularly when the mould moves from platen to platen during the cycle, heated platens or back plates are the most convenient form of mould heating but do not necessarily give the best temperature distribution. The temperature distribution may be corrected to a degree in certain shapes of moulds by the use of induction heating.

In some cases the selection of cycle pattern has an effect on the processing conditions. One case in which this may become apparent is in the regulation of the injection nozzle or front zone temperature. When the nozzle is retracted from the moulding at the beginning of the cure time, the heated mould exerts a minimum of influence on the nozzle temperature. However, this cannot be the invariable cycle for the machine as operation in this way is only possible if the rubber

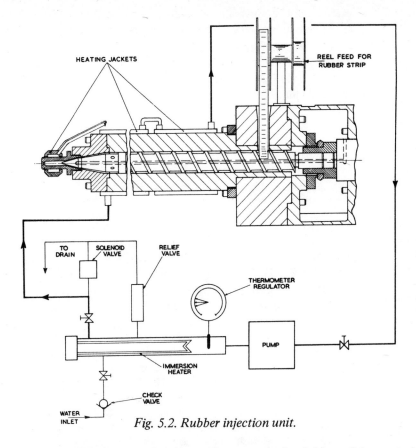

Fig. 5.2. Rubber injection unit.

precharging can take place without nozzle drool. Most of the controls
which are effective in the precharging phase have their effect intensified
if the rotation of the screw takes place at the end of the cure time,
as closely as possible before the injection stroke for which the
precharge is made.

The screw delay function has been shown to be a major factor in
establishing rubber temperature-control by precharge adjustments.

In conclusion therefore, it is possible to say that the injection moulding
machine possesses many more controls than a simple compression
(Fig. 5.3 Plate 8). When these controls are divided into functional
groups they are, however, less confusing and their operation is made
easier to understand.

DISCUSSION

Dr L. J. GERHARDT *(Vitamol Precision)*: Are separate runner plates
often used?

ANSWER: Cold runner moulds appear to be designed on the same
principle as hot runner moulds for thermoplastics. A separate runner
plate is a necessity as this must be controlled at a different temperature
from that used in the mould cavities. For rubber work this cold runner

plate should be easily opened up for extraction of the rubber contained in the runners when the moulding run is concluded.

F. SCOGNAMIGLIO *(G. Angus and Co. Ltd)*: Are there any means of selecting and reading the locking force of injection moulding machines?

A: It is sometimes necessary to regulate the locking force of injection moulding machines: first, to protect small or soft moulds against excessive loads and secondly, to provide a 'squeeze' action which can be programmed with the injection unit controls to provide certain special affects.

Simple reduction of locking force is most easily achieved in a direct hydraulic injection moulding machine when it depends only on the setting of an appropriate relief valve. In the case of toggle machines, the mould height adjustment is used to regulate the position at which the toggle mechanism starts to take up its load and hence, the final strain and locking force achieved is not normally provided on standard machines. However, the measurement of tie bar stretch by dial gauge or strain gauge is a relatively simple matter and this provides a direct indication of locking force after an elementary calculation.

F. J. BLOWERS *(George Angus)*: Please enlarge upon the parameters considered in determining screw length.

A: The length of screw used for processing rubber has been determined empirically as a result of development experience. Current practice in thermoplastic injection moulding indicates that a length: diameter ratio of 16:1 is an optimum. In rubber processing rather more varied opinions exist but a length:diameter ratio of 10–12:1 has a general acceptance. Screw design for rubber is arranged to provide no compression in flight depth and the screw length needed is in order to provide an adequate transfer area so that steady, gentle, well distributed heating may be arranged rather than to accommodate the more marked change of physical state found in thermoplastic work. It is worth noting that a machine with a relatively short injection stroke will lose less of its screw processing length when the screw is fully retracted. For a given shot displacement its screw diameter will be larger than that of a longer stroke machine and it will therefore have the advantage of a larger heat transfer area which varies less during each cycle. Designs which achieve their shot displacement by relatively small diameter screws with long injection strokes, have a more variable heat transfer area available during each cycle. A longer length:diameter ratio can provide a small diameter screw with adequate heat transfer area but is less effective in reducing the variability due to its stroke. The advantage appears to be with screws of length:diameter ratio of 10–12:1 with a stroke/bore ratio of not more than 2:1.

W. S. PENN *(Borough Polytechnic)*: 1. Are separate controllers used to control cooling of screw and barrel? 2. To what extent is cold runner moulding employed?

A: *Question 1*; Barrel cooling is normally indirect and achieved through the same fluid heat transfer medium utilized for heating. For experimental purposes machines have been equipped with 'crash' cooling arrangements whereby cold or refrigerated water can be introduced in liberal quantities to the barrel jackets or to a heat exchanger in the fluid circulating system. This arrangement is not normally required in production moulding and indeed, is not desirable, as its undisciplined use could lead to temperature shock failures in machine parts. It is normally more satisfactory to terminate a moulding run by purging with unvulcanized rubber.

A: *Question 2*; Cold runner moulding is not employed extensively at present but its use will become more apparent as the economic advantage of eliminating the waste material in screws and runners becomes apparent. It is not limited to multi-cavity work and can be arranged to feed a single cavity via a temperature controlled reservoir or 'reversed screw'

CHAPTER **6**

Cure systems for efficient injection moulding

D. A. HAMMERSLEY, M.A. (Cantab.)
Monsanto Chemicals Ltd

6.1 INTRODUCTION

THE high temperatures employed in the injection moulding process put special demands on the curing system and it is necessary to ensure that the ones employed allow the injection machine to be used at its most efficient operating conditions. This chapter is concerned with the special requirements of curing systems for injection moulding and with their design in meeting these requirements.

6.2 TECHNICAL REQUIREMENTS OF CURING SYSTEMS

An overall view of the technical properties of a curing system can be obtained by plotting some physical property of the rubber (e.g. modulus) against time at a fixed temperature. A typical curve is shown in Fig. 6.1. The scorch time is the length of the initial delay period before the modulus begins to rise. It can be defined as the time to a 5 point rise in Mooney viscosity (t_5) or a 2in. lb rise in torque on the Monsanto oscillating disc rheometer (t_5). The cure rate is shown by the slope of the steeply rising part of the curve and the optimum cure time is defined as the time to reach 90% of the maximum modulus (T 90%). Ideally, the modulus curve should flatten to a good plateau but in some cases continues to rise or starts to fall. This continued rise is called marching modulus and the fall, reversion. Marching modulus is particularly prevalent in nitrile rubbers and reversion in natural rubber and cis-polyisoprene. The level of cure is shown by the height of the plateau—that is the modulus of the final product.

In the injection moulding process heat is put into the compound progressively by transfer from the barrel, by mechanical work in the nozzle and finally by transfer from the mould surface. The scorch time must be sufficient for the compound to remain plastic until mould flow is complete and then the cure rate should be as fast as possible. Faster cure rates give direct improvements in throughput and indirectly longer scorch times also lead to an improvement since they enable higher barrel temperatures to be used, thus decreasing warm-up times in the mould. Inadequate scorch time for a given injection system will cause poor surface finish, insufficient flow at the mould profiles, chattering at the injection pip and, in severe cases, scorch in the nozzle.

It is particularly important in injection moulding that the compound used maintains a uniform modulus (or flat plateau) once cure has occurred. At the very high temperatures used, the rate of cure of thick sections will be strongly influenced by the rate of heat transfer through the rubber. Overcure of the surface relative to the centre may occur if the modulus varies with time of cure and thus can lead to variable properties. It is also necessary to avoid curing systems which can lead to reversion in natural rubber and cis-polyisoprene because the tendency to this form of degradation is accentuated by high temperatures.

Ideally, curing systems in injection moulding must give adequate scorch time, fast cures, appropriate physical properties and plateau cures.

6.3 EFFECT OF TEMPERATURE ON CURING PROPERTIES

Before the advent of injection moulding most work on the development of curing systems was done using processing temperatures in the range $90°-120°C$ and curing temperatures in the range $120°-160°C$. The effects of high temperatures ($>160°C$) on the relative merits of various curing systems will therefore be considered.

Cure curves produced by the Monsanto rheometer for four different accelerators in a black NR compound at $140°C$ are shown in Fig. 6.2 and at $180°C$ with an expanded time scale in Fig. 6.3. Differences in scorch time, rate of cure and final modulus between the different accelerators are seen to be reproduced in the same relative order at the two temperatures.

Table 6.1 shows the scorch times for seven accelerators in the same compound at three temperatures. The pattern at $121°C$ is repeated at $140°C$ and again at $180°C$ although values are closer together.

It can be concluded that differences between one accelerator system and another at any one temperature are repeated over the entire temperature range normally used for curing solid rubber.

Table 6.1 Effect of temperature on scorch time

Accelerator at 0·5 phr	Scorch time rheometer t_2		
	at 121°C mins	at 140°C mins	at 180°C mins
Ethasan (ZDC)	5½	3	70
Thiurad (TMTD)	9	4	90
Thiotax (MBT)	15	5½	105
Thiofide (MBTS)	18½	7	125
Santocure (CBS)	29	10	135
Santocure MOR	33	11	145

Base Stock:	RSS1	100
	HAF black	50
	Process oil	3
	Zinc oxide	5
	Stearic acid	3
	Sulphur	2·5
	Accelerator	0·5

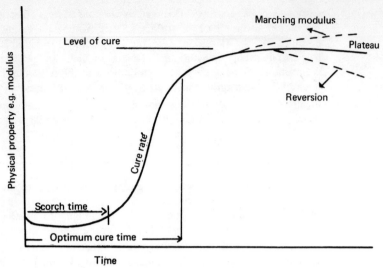

Fig. 6.1. Cure curve.

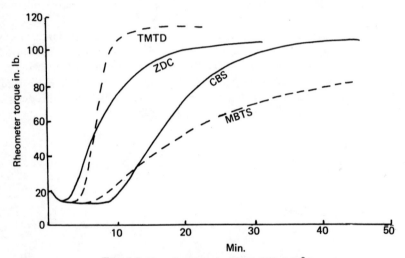

Fig. 6.2. Accelerators in black NR 140°C.

6.4 CURING SYSTEMS IN COMMON USE

A manufacturer transferring from a compression to an injection moulding process may fairly safely carry out the first trials without modifying the compound, relying on adjustments of barrel temperature to obtain reasonable operating conditions.

Some typical formulations for different polymers are listed in Tables 6.2–6.6 together with curing systems selected to offer a range of processing and cure requirements. These are based either on MBTS, sulphenamides or Sulfasan R because of the need for a certain minimum of scorch safety in the compounds.

48

Table 6.2 Black NR formulations

Base Mix:	RSS4	70
	Whole tyre reclaim	60
	MPC black	75
	Dutrex R	10
	Zinc oxide	40
	Stearic acid	5
	Paraffin wax	2
	Santoflex 13	1

Curing Systems:	A	B	C	D	E
Sulphur	2·5	2·5	2·5	–	–
Sulfasan R	–	–	–	1·4	1·2
Santocure	1·2	1·2	–	1·4	1·2
Santocure MOR	–	–	1·2	–	–
Thiurad	0·3	–	–	0·2	0·5

A, B, and C are suitable for thin section products and are in ascending order of scorch time. D and E are efficient vulcanizing systems suitable for thick sections. They give much reduced reversion and improved ageing resistance.

Table 6.3 White NR formulations

Base Mix:	Pale crepe	100
	Whiting	75
	China clay	25
	Zinc oxide	4
	Stearic acid	1·5
	Paraffin wax	2
	Santowhite FM	2

Curing Systems:	A	B	C	D
Sulphur	2·5	2·5	2·5	–
Sulfasan R	–	–		1·4
Thiofide	1·5	1·5	2·0	1·4
Monothiurad	–	0·5	–	–
Thiurad	0·5	–	–	1·2

Table 6.4 SBR formulations

Base Mix:	SBR 1500	100
	SRF black	75
	Process oil	8
	Zinc oxide	4
	Stearic acid	2
	Sulphur	2

Accelerator Systems:	A	B	C	D
Thiofide	2·0	2·0	—	—
Santocure	—	—	1·5	2·0
Thiurad	0·5	—	1·0	0·5
Monothiurad	—	0·5	—	—

SBR/BR blends can be used in otherwise unaltered formulations.

Table 6.5 Nitrile rubber formulations (NBR)

Base Mix:	Nitrile 1002	100
	SRF black	80
	Dioctyl phthalate	5
	Zinc oxide	5
	Stearic acid	1

Curing Systems:	A	B	C	D
Sulphur	1·5	1·5	0·5	—
Sulfasan R	—	—	—	—
Thiurad	0·5	—	3·0	—
Thiofide	1·0	1·5	—	3·0

A and B are conventional curing systems which may be adequate where ageing resistance is not a particular problem. C is a low sulphur system giving much improved ageing but its scorch time is usually sufficient only for ram type injection. D combines excellent ageing with a scorch time long enough for most applications.

6.5 ADJUSTMENT OF CURE SYSTEMS

In many cases the first curing system tried will be found to need some modification to meet the precise requirements of a given process. The method of adjustment can be seen by reference to Fig. 6.4, which shows the cure curves for four common types of accelerator. The one required for a particular compound is determined by the scorch time needed and intermediate times are usually met by combinations. It will be observed that most of the curing systems listed in the previous section, 6.4 consist of combinations of MBT, MBTS or Santocure with TMTD or TMTMS.

Table 6.6 Natural rubber—white stock

	Base Stock:	Pale crepe	100
		Zinc oxide	5
		Whiting	80
		Stearic acid	0·5

Properties	Vulcanizing/antioxidant system			
	Conventional		Efficient	
	Antioxidant*	1·5	(No added antioxidant)	
	Sulphur	1·5	DTM	1·0
	MBT	1·0	MBTS	1·0
	DPG	0·5	TMTD	1·5
Mooney scorch 121°C., t_5, min.	14		13	
Optimum (90%) cure 148°C, min.	17		18 ·	
Tensile strength kg/cm².	226		214	
Modulus 300% kg/cm².	50		52	
Elongation at break %	580		570	

Percentage retention of tensile strength in oven at 90°C

days		
8	69	77
18	58	74

*Antioxidant—4,4'Thiobis(2—tertiary butyl-5-methyl-phenol).

It will also be observed from Fig. 6.4 that MBTS gives a slower cure than either TMTD or Santocure NS and it might be concluded that scorch times equivalent to that of MBTS are better obtained by combining Santocure NS with TMTD. Fig. 6.5 shows this to be true and also shows that activating MBTS with another widely used material, DPG, gives a poorer rate than an NS/TMTD system.

Thus if Santocure NS/TMTD combinations are used scorch time may be varied by adjustment of the NS/TMTD ratio and in each case cures are faster than with other systems.

6.6 EFFICIENT VULCANIZING SYSTEMS

Efficient vulcanizing (EV) systems are defined as those where a high proportion of the sulphur is used for crosslinking purpose. These systems have two main advantages over conventional systems, giving vulcanizates with reduced reversion and better ageing characteristics. In addition to these advantages, EV systems based on Sulfasan R (dithiodimorpholine—DTM) are very versatile, enabling a wide range of scorch times, cure rates and states of cure to be chosen at will.

For injection moulding of thick sections it is particularly important to avoid reversion and EV systems give the complete answer to this problem. Fig. 6.6 illustrates this for a white NR stock. The conventional

51

system (sulphur/MBTS/DPG) shows reversion immediately after the maximum modulus is reached, whereas the EV system (DTM/MBTS/TMTD shows no reversion even after three times the optimum cure time. EV systems can be developed to give equivalent cure propoerties with much improved ageing as compared with a conventional cure, even when antioxidants are omitted, as shown in Table 6.6.

The design of DTM based EV systems and their great versatility are illustrated in a black NR stock in Table 6.7. A conventional sulphur/CBS system is first replaced by equal proportions of DTM and CBS. This gives equivalent product properties but very much longer scorch and cure times. Shorter scorch and cure times are obtained by the progressive addition of TMTD as a regulator with corresponding reduction in CBS and DTM levels. The third EV system gives both longer scorch time and faster cure than the conventional system—a further advantage.

Similar principles apply in SBR where, although reversion is not normally a problem, EV systems are used because of their very good ageing characteristics and great versatility in adjustment of scorch time, cure rate and state of cure. The variations of scorch time, modulus and cure time with curative levels are shown in Figs 6.7, 6.8 and 6.9.

Table 6.7 Use of sulphenamide/dithiomorpholine/TMTD system to match stress/strain properties of conventional cure system

Base Stock: NR	100
HAF black	50
Zinc oxide	5
Stearic acid	1
Oil	3
IPPD	1·0

Properties	Conventional			Vulcanizing system Mono/disulphidic			
				1	2	3	4
	Sulphur	2·5	CBS	1·5	1·4	1·3	1·0
	CBS	0·5	DTM	1·5	1·4	1·3	1·0
			TMTD	0	0·2	1·3	1·0
Mooney scorch 120°C., t₅, min.	24			52	35	28	16
Optimum (90%) cure 140°C., min.	24			40	24	22	16
Tensile strength kg/cm².	278			280	275	280	280
Modulus 300% kg/cm².	130			125	125	130	135

6.7 EFFECT ON CURE OF VARYING LOADINGS OF COMPOUNDING INGREDIENTS

Most compounding ingredients have some effect on curing properties. When the loadings of fillers or oils are varied to meet differing product specifications it may be necessary to adjust the curing system. Similarly changed specifications may often be met by changed curatives.

This section, therefore, summarizes the main effects on cure properties of varying loadings of curatives, fillers and oils.

Increasing accelerator loading gives increased rate of cure and increased modulus. Changes in scorch time are usually small with sulphenamide accelerators and may be in either direction. This is illustrated in Fig. 6.10. Increased hardness and decreased compression set are also obtained by increasing accelerator loading.

Increasing sulphur level gives increased modulus and slightly increased rate of cure (Fig. 6.11) but generally worsens ageing properties.

Increasing the loading of stearic acid gives increased vulcanizate modulus but decreased initial viscosity (Fig. 6.12) while increased oil level gives decreased modulus and decreased viscosity (Fig. 6.13). It is sometimes possible to cheapen a compound by more oil extension and compensating for the lower modulus by increasing the accelerator loading.

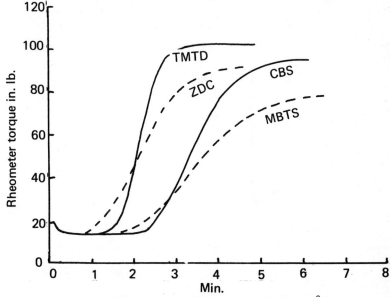

Fig. 6.3. Accelerators in black NR 180°C.

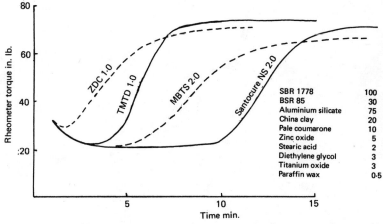

Fig. 6.4. Accelerators in resin reinforced SBR 153°C.

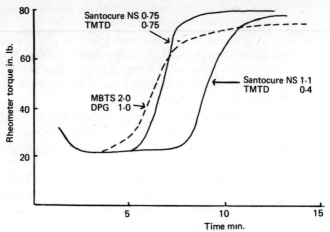

Fig. 6.5. *Accelerators in resin reinforced SBR 153°C.*

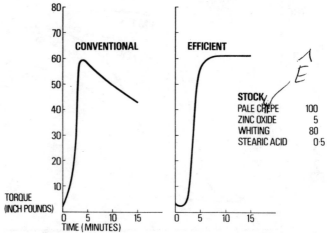

Fig. 6.6. *Comparison of rheometer curves 180°C.*

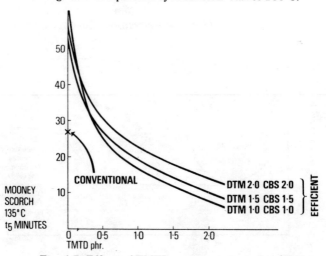

Fig. 6.7. *Effect of TMTD on processing safety/SBR stock.*

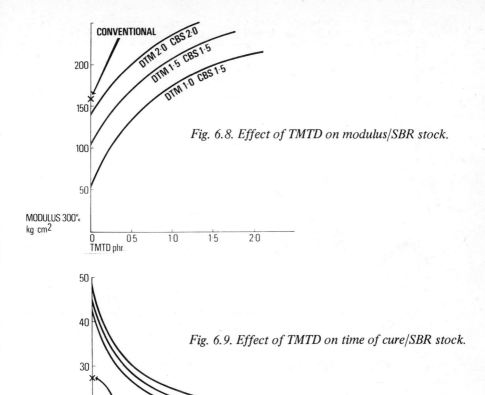

Fig. 6.8. Effect of TMTD on modulus/SBR stock.

Fig. 6.9. Effect of TMTD on time of cure/SBR stock.

With synthetic polymers which are available in grades of differing unsaturation the highest unsaturation grades give the shortest scorch times, the fastest cures and the highest moduli. This is illustrated for butyl rubbers in Fig. 6.14. Where a change to a lower unsaturation grade is necessitated by tighter ageing specifications the effect on cure may be counteracted by increased accelerator loading.

Increasing carbon black loading usually gives decreased scorch time and increased modulus (Fig. 6.15). When necessary decreased scorch time may be counteracted by a change in accelerator system, e.g. a change from Santocure to Santocure MOR or increased sulphenamide ratio with a sulphenamide/TMTD combination.

6.8 SUMMARY AND CONCLUSIONS

Accelerator systems for injection moulding should be chosen to give adequate scorch time, fast cure without reversion and appropriate product properties.

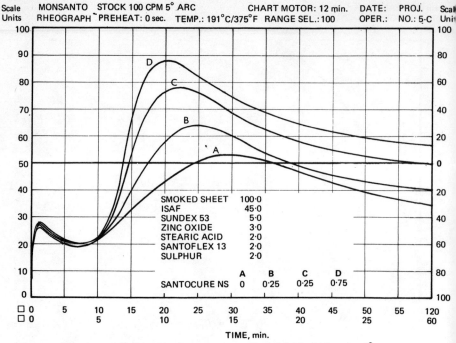

SMOKED SHEET	100·0
ISAF	45·0
SUNDEX 53	5·0
ZINC OXIDE	3·0
STEARIC ACID	2·0
SANTOFLEX 13	2·0
SULPHUR	2·0

	A	B	C	D
SANTOCURE NS	0	0·25	0·25	0·75

TIME, min.

Fig. 6.10. NR Levels of accelerator. Model 100. 100 cpm, 5° arc.

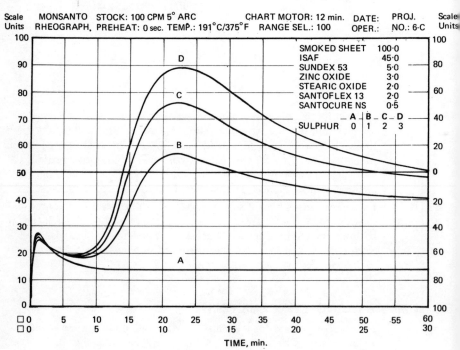

SMOKED SHEET	100·0
ISAF	45·0
SUNDEX 53	5·0
ZINC OXIDE	3·0
STEARIC OXIDE	2·0
SANTOFLEX 13	2·0
SANTOCURE NS	0·5

	A	B	C	D
SULPHUR	0	1	2	3

TIME, min.

Fig. 6.11. NR Levels of sulphur. Model 100. 100 cpm, 5° arc.

56

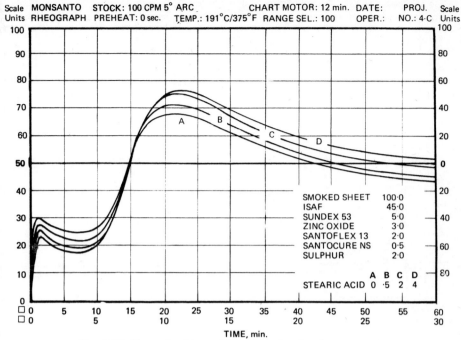

Fig. 6.12. Natural rubber with varying levels of stearic acid.

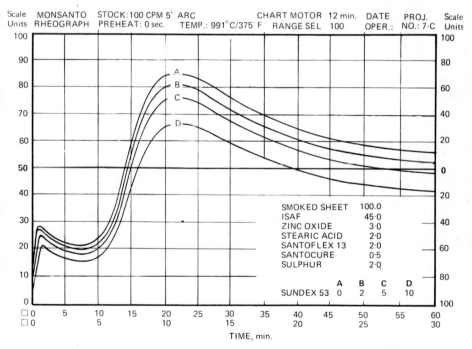

Fig. 6.13. NR Levels of processing oil. Model 100. 100 cpm, 5° arc.

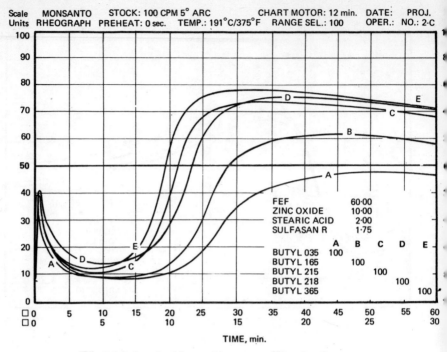

Fig. 6.14. Butyl with varying types of butyl polymer.

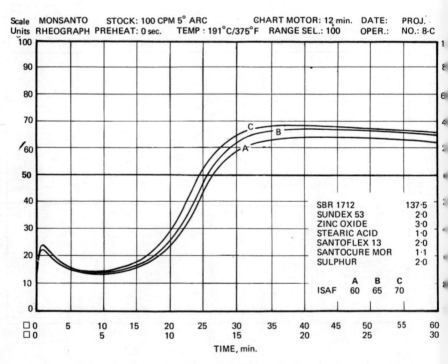

Fig. 6.15. SBR Levels of ISAF carbon black. Model 100. 100 cpm, 5° arc.

58

When moulding thick sections from polymers which revert (e.g. NR and cis-polyisoprene) EV systems should be used to minimize reversion. Combinations of a sulphenamide (e.g. Santocure), Sulfasan R (dithiodimorpholine) and Thiurad (TMTD) are ideal, and the ratios can be varied to meet precise machine operating conditions and product requirements.

Where reversion is not a problem conventional sulphur/accelerator systems can be used and the following accelerators will give the best cure rates for each scorch time requirement.

Santocure MOR   Decreasing
Santocure NS   scorch
Santocure NS/thiurad (TMTD) time

Accelerator loadings may be increased to give improved product properties or to counter the effect of oil addition. The effect on scorch of increased black loading may be countered by a change to a longer scorch accelerator system.

A number of the trade names mentioned in this paper, including Santocure, Thiofide and Santoflex, are registered trade marks.

DISCUSSION

R. F. POWELL *(Trist Mouldings and Seals)*: How can scorch times be directly correlated with machine operation and can this be used as a control?

ANSWER: A knowledge of the scorch times required for particular process conditions can only be obtained by actual operating experience. The scorch time and the other parameters of the cure curve can be used for process control and the Monsanto oscillating disc rheometer is particularly valuable for this purpose.

DR L. GERHARDT *(Vitamol Precision)*: 1. Do you get the same efficiency if you blend polymers first and ingredients afterwards, than if the ingredients are dispersed first in polymers in which they are not as soluble? 2. Are shear rates of the Monsanto rheometer high enough to compare with injection shear rates?

A· *Question 1*; The usual practice is to blend polymers before the addition of curatives. In some cases Dr Gerhardt may be right but diffusion before cure would reduce the advantage. Dr Wheelans added that NRPRA had found that diffusion was so fast that the equilibrium would be reached well before cure.

A: *Question 2*; Shear rates obtainable in the Monsanto oscillating disc rheometer are between 3½ cpm and 750 cpm. This is considerably less than the shear rates occurring in injection moulding but rheometer cure curves are nonetheless valuable for predicting the behaviour of the curing system.

E. GRINDLE *(Dowty Seals)*: In cheap compounds using high proportion of oil or white fillers what steps are required to offset adsorption of the cure system at high temperature?

A: Higher loadings of accelerators are usually used in cheap white filled compounds. There seems to be little difference in this respect between the various accelerator combinations evaluated.

B. WHITTAKER *(RAPRA)*: Are the 'efficient' curing systems still efficient in blends of polymers at injection moulding temperatures or are the rates of cure and states of cure so different that one of the polymers may be starved to curatives?

A: Sulfasan R curing systems *do* give good ageing and low reversion in many polymer blends.

CHAPTER 7

Carbon black in injection moulding compounds

N. C. H. HUMPHREYS, F.I.R.I.
Philblack Limited

7.1 INTRODUCTION

IN common with the majority of practical advances in the manufacture of rubber articles, injection moulding has developed empirically, with scientific explanations for the results obtained lagging far behind, and theoretical forecasts for the behaviour of a given compound in a new injection moulded application being of little direct value. This is no reflection upon the considerable volume of published work about theoretical aspects of the process but rather illustrates the immense complexity of the subject.

Were rubber a Newtonian liquid with simple heat transfer properties, and were vulcanizing systems subject to simple interpretation in terms of temperature/time behaviour, the subject would become much easier. In the plastics industry, where injection moulding has long been an established method, these factors are of less importance, or do not apply at all, although plastics technologists are not without their own problems, such as the need to make thermosetting resins flow along hot mould surfaces. Rubber, under which general title all the common vulcanizable elastomers may be classified in the present context, in fact presents so complex a picture in rheology that the scientific forecasting of the behaviour of new compounds or new moulds is a largely unconquered field.

7.2 PLASTICITY FACTORS

It is against this involved background that the study of carbon black in compounds for injection moulding must be made. This paper is not intended to reveal new facts nor to provide ready-made formulae for specific purposes, but rather to discuss and integrate ideas.

To begin with, carbon blacks add a further factor to the already complicated picture because they form physical and chemical linkages both with polymers and between individual black particles. The nature and strengths of these linkages have not yet been fully explained, although their overall effects are, in general, fairly well understood. For example, measurements made in the three basic types of plasticity testing machines, i.e. parallel plate, extrusion and shearing plastimeters, bear little relationship to one another when a compound containing a reasonable quantity of reinforcing carbon black is under examination. This is largely because the shearing plastimeter breaks down much of the 'structure', the extrusion plastimeter breaks down less of the 'structure' and the parallel plate plastimeter has little effect on 'structure

Plasticity control of compounds intended for injection moulding is desirable and the shearing type of plastimeter is probably the best instrument to use. However, plasticity control is not so important as might be imagined because heat build-up at the nozzle depends primarily on work done, which is dictated by the pressure exerted by the machine. Build-up of pressure occurs over a finite time, which has a considerable effect in maintaining equilibrium of the temperature at the nozzle over a range of plasticities. Therefore, plasticity variations from batch to batch of the same nominal compound will not necessarily result in scorching because of heat build-up at the nozzle.

Injection times might also be expected to vary profoundly with changes in plasticity but, here again, it is found in practice that the variation in time is relatively small. G. W. Morris and D. A. W. Izod [1] demonstrated difference of only 5 sec. in injection time over a range of HAF loaded compounds in various base polymers, the loadings varying from 25–60 phr and the Mooney viscosities from 30–90.

From the foregoing data it can be deduced that widely varying carbon black loadings can be used on a practical basis for injection moulding and, quite possibly, despite the higher temperatures employed, scorching danger may often be less than exists in conventional moulding.

The Monsanto rheometer may be a useful means of determining the characteristics of stocks for injection moulding and some curves for carbon black stocks are given in Fig. 7.1.

7.3 IMPROVED PHYSICAL PROPERTIES

For carbon black loaded compounds the screw ram type of injection moulding machine is probably the best available so far because the compound is simply strip fed in as for a normal extruder, is mechanically worked by the screw, and it is brought to an even temperature just prior to the retracted screw coming forward to act as an injection piston. The action of the screw and the work done at the injection nozzle in blending the rubber and thus giving better dispersion is thought by some workers [2] to result in a better level of physical properties in vulcanizates compared with results of conventional moulding, but the author feels this may not be entirely correct. Certainly a badly mixed carbon black reinforced compound will be improved by subsequent extrusion, but there is evidence that a slight but significant improvement is often obtained over conventional moulding when well mixed compounds are injection moulded.

7.4 HEAT CONDUCTIVITY

In the opinion of the author there is a need for deeper studies of heat conductivity of compounds because this could well be the clue to some differences in physical properties between the same stock when press moulded and when injection moulded. Of course, this type of comparison is not clear cut because injection moulding temperatures are higher and times shorter. Another aspect is that there must be considerable variations in the temperatures attained at the centre of a

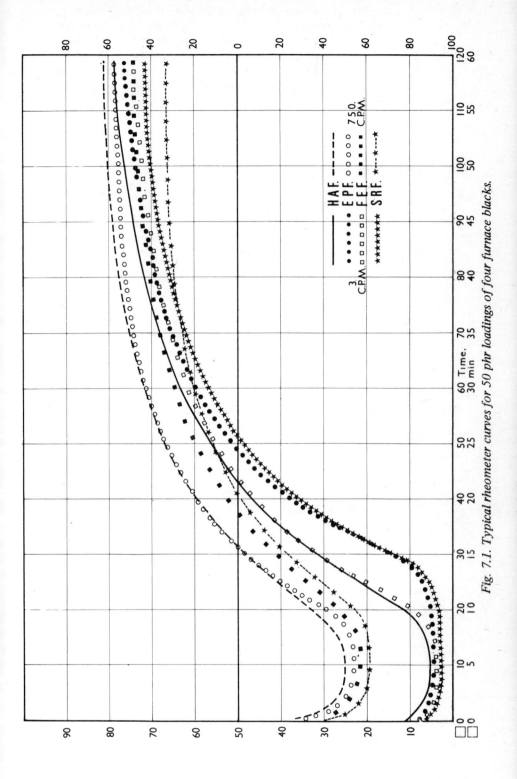

Fig. 7.1. *Typical rheometer curves for 50 phr loadings of four furnace blacks.*

compound loaded with carbon black and a compound loaded with a similar volume of lower heat conducting white filler. The high heat conductivity of carbon black makes it a desirable compounding ingredient to use in the injection moulding process.

This whole question of heat transfer is closely associated with curative system behaviour, including reversion, which will not be discussed here beyond stating that carbon black loaded compounds for injection moulding are frequently similar to, or identical with, compounds previously used to produce the same article in a press.

7.5 TYPICAL COMPOUNDS

Compounds containing substantial loadings of carbon black are normally used in high performance vulcanizates. These compounds may be based on natural rubber, SBR, or one of the special purpose polymers. Examples of compounds which have been successfully used in both conventional and injection moulds are as follows:

Natural rubber	100
CBS	0·5
Zinc oxide	5
Plasticizer	10
HAF black	55
Stearic acid	1·5
Sulphur	2·5
Butyl rubber	100
Stearic acid	0·5
MBTS	0·5
ZDC	2
Zinc oxide	5
Plasticizer	12
FEF black	60
Sulphur	2
Nitrile rubber	100
Stearic acid	1
TMTMS	0·6
FEF black	30
HAF black	30
Plasticizer	10
Sulphur	1
Zinc oxide	5

7.6 TRIMMING AND OTHER PROBLEMS

When deciding between conventional and injection moulding the economic considerations can be involved and will depend on such factors as numbers to be produced in relation to machinery costs, comparative direct labour and overhead charges, quality of finished product, reject rate, etc.

Since it is the case that carbon black loaded vulcanizates are usually of a high degree of toughness, trimming is often difficult and expensive.

Hence, injection moulding, which frequently produces less flash because of mould designs which are not possible for press moulding, is an attractive proposition. This reduction in trimming difficulty is not only valuable in saving labour but, for example, in precision mouldings such as oil seals, may make a considerable contribution towards quality of article and reduction of trimming fault rejects.

At an earlier stage of the production process the injection of a compound can often help to reduce rejects arising from poor joining of the rubber in a complicated mould, such as a bellows shaped gaiter, of which an excellent example is to be found in the literature [3].

Bonding of rubber to metal usually involves highly reinforced compounds containing carbon black. The problems in bonding are numerous and complicated but injection moulding frequently offers the possibilities of (a) presenting a clean rubber surface to the metal, (b) obtaining optimum compound flow before vulcanization has reached a stage when the bond will be adversely affected, (c) obtaining positive pressure over all parts of the bonding interface during vulcanization.

7.7 SHRINKAGE

A matter which must not be overlooked when deciding upon the relative merits of conventional and injection moulding of precision articles is the question of shrinkage of the rubber. Carbon black in itself is not the basic cause of dimensional instability, but precision mouldings very often need the reinforcement which only carbon black can impart, and the advantages in having less flash on oil seals, referred to earlier in this paper, must not be allowed to blind the designer to the dangers of constructing a very expensive mould, or set of moulds, for an injection machine which, because of inadequate knowledge of the shrinkage behaviour of the compound, turn out to be useless. It is quite common practice to make conventional moulds to the lowest forecast dimensional tolerances and to enlarge them if trials in the factory show the desirability of so doing; it is not always so easy to do this with injection moulds and, furthermore, the skrinkage behaviour of a compound which has been in use in a press mould may not be at all the same if a change is made to injection moulding.

7.8 CONCLUSIONS

The modern furnace black manufacturer has techniques to vary his products in numerous ways, of which particle size and 'structure' are among the more important, and it may well be that blacks will have to be developed especially for injection moulding applications at some time in the future. For the present it seems that the normal types of carbon black, with their comparitively small effect on injection times, their ability to process with little or no change from conventional vulcanizing systems and their high heat conductivity, fit very well into the modern injection moulding picture.

ACKNOWLEDGEMENTS

The author's thanks are due to the Empire Rubber Company for permission to illustrate a moulding and to Mr D. A. W. Izod of the Rubber and Plastics Research Association of Great Britain for helpful discussions.

REFERENCES

1 Morris, G. W. and Izod, D. A. W., Injection Moulding of Natural and Synthetic Rubbers. *Rubber and Plastics Journal* Feb. 1965, **46**, 2.

2 Watson, W. F., Injection Moulding of Rubber. ACS (Rubber Division) Meeting. New York. Sept. 1966.

·3 Rubber Injection in operation at Empire Rubber. *Rubber Journal* July 1965, **147**, 7.

DISCUSSION

W. S. PENN *(Borough Polytechnic)*: Since viscosity does not appear to be a criterion of injectability, would you think that 'scorch' is more appropriate?

ANSWER: Scorch is the essential criterion but viscosity plays a part in scorch since differences in viscosity lead to different degrees of heat generation, which alters the scorch property.

QUESTION: Is the Monsanto rheometer the best compromise instrument?

A: The Monsanto rheometer is, in the opinion of the author, the best compromise instrument available.

R. F POWELL *(Trist Mouldings and Seals)*: Can Mooney viscosity be used as a control on the IM process?

A: Mooney viscosity can be used as a control in many injection moulding operations but only in the light of experience built up on a given compound and set of operating conditions. In the experience of the author it is not usually possible to forecast the behaviour of a new compound or an existing compound in a new set of injection conditions from either a Mooney figure or any other plasticity measurement at present available.

CHAPTER 8

Factice in injection moulding compounds

C. FALCONER FLINT, Ph.D., D.I.C., B.Sc., A.R.C.S., F.R.I.C., F.I.R.I.
and H. ROEBUCK, A.N.C.R.T., A.I.R.I.
Factice Research and Development Association

8.1 INTRODUCTION

The relatively modern name factice marks the improvements resulting from the application of research and development work to the friable solid made by vulcanizing triglyceride oils with sulphur. When manufactured in the United Kingdom, this rubber compounding ingredient is normally graded as follows on the basis of hot acetone extract value:

Grade 1 factice—acetone extract less than 20%
Grade 2 factice—acetone extract 20%–35%
Grade 3 factice—acetone extract greater than 35%

Factice has long been used in the rubber industry as an aid to processing and productivity, proving particularly beneficial in such application as extrusion and calendering. It has also been shown by Flint [1] and by Flint, Wardle and Clark [2] that factice is capable of activating the curing system.

Given the desirability, or the necessity in some cases, of using factice as a processing aid in a rubber compound, it is important to know how the compound would then behave in the injection moulding process, where shear rates can be 1 000 times higher than those obtained on conventional rubber processing equipment, and where vulcanizing temperatures are higher than those applied in normal compression moulding. In this process, the high shear rates result in very short mould-filling times, and it might be thought that any further improvement in flow rate was scarcely important. The ensuing work gives a quantitative assessment of the effect of factice in shortening the mould-filling time, but also demonstrates a more important contribution to productivity in that, under injection moulding conditions, factice causes sufficient set up in the early stages of the cure to reduce porosity and thus reduce the minimum time required to permit the article to be removed from the mould without distortion.

8.2 ANKERWERK MACHINE TRIALS

Preliminary investigations into the effect of factice on compounds to be processed by injection moulding were carried out on an Ankerwerk V28-140 screw plunger (i.e. a reciprocating screw) type machine with the kind co-operation of Hamilton Machinery Sales Ltd. Two compounds based on polyisoprene and SBR 1509, taken from trade literature and recommended for injection moulding, were investigated

The photographs, Figs 8.1, 2 and 3 will be found on Plate 9, (facing page 70).

with and without 10 phr of a Grade 1 factice, and 10 phr of a Grade 3 factice (see Table 8.1). The results of these trials are summarized in Table 8.2.

Table 8.1

Polyisoprene		*SBR*	
Isoprene 305	100	SBR 1509	100
Sulphur	1·5	Mineral oil	20
Zinc oxide	10	FEF black	30
Stearic acid	2	Fortafil A70	50
SRF black	20	China clay	20
EPC	35	Zinc oxide	5·0
ISAF black	15	Pale coumarone resin	2·5
Gravex 21 oil	8	Paraffin wax	0·5
Heliozone	1	Stearic acid	2·0
Antioxidant 2246	1	Diethylene glycol	3·0
Manosperse A	3	CBS	1·0
CBS	1	MBTS	1·0
TMT	0·2	TMT	0·5
Grade 1 factice	0 and 10	Sulphur	2·0
Grade 3 factice	0 and 10	Grade 1 factice	0 and 10
		Grade 3 factice	0 and 10

Table 8.2 Results of trials on Ankerwerk machine

Polyisoprene compounds

	No factice	10 phr Grade 1 factice	10 phr Grade 3 factice
Mould filling time (sec.) at 11 000 psi	1·6	1·7	1·7
Time to 'min. prac. cure' (sec.) at 200°C	60	45	30
Temperature after ejection (°C)	89	93	95

SBR compounds

	No factice	10 phr Grade 1 factice	10 phr Grade 3 factice
Mould filling time (sec.) at 11 000 psi	1·7	1·5	1·4
Time to 'min. prac. cure' (sec.) at 200°C	45	30	30
Temperature after ejection (°C)*	105	112	105

* i.e. temperature of compound after ejection from barrel of machine.

The effect of factice was to reduce cycle time by as much as 40% when cure time was taken as the time required to produce a moulding which was free from porosity and could be demoulded without distortion. This was termed 'minimum practical cure'. Factice played a major part in reducing cure time whilst having little effect on injection time at realistic injection pressures.

The conditions of operation of the Ankerwerk machine were as follows:

(i) Barrel temperature 70°C
(ii) Mould temperature 200°C
(iii) Pressure held for 25 sec. after ejection
(iv) Back pressure 100 psi

The temperature build-up on ejection was measured by consolidating the rubber as it was ejected, and then observing the temperature by means of a needle thermocouple. The temperature rise in the factice compound was in all cases greater than or equal to the temperature rise observed in the no factice compound. The temperature rise attributable to factice seemed to depend both on the compound and the type of factice used.

It has been suggested by other workers [3] that the cure time is influenced by temperature rise on ejection and is dependent on the heat history of the compound prior to moulding [4] Although temperature rise on ejection undoubtedly influences the rate of cure, this cannot be the full explanation for reduction in cure time caused by factice. The SBR compound containing 10 phr of the Grade 3 factice had a shorter cure time than the no factice compound, some other explanation for the effect of factice in reducing time to minimum practical cure is therefore required.

The physical test data shown in Table 8.3 demonstrate the ability of

Table 8.3 Physical properties determined on sections of the Ankerwerk injection moulding samples.

Compound	Cure time at 200°C in sec.	BSH	M.300 (psi)	TS (psi)	E @ B (%)	Indentation hardness* (mm. $\times 10^{-2}$)
IR.305 NF	45			Not cured		
	60	61	1 100	2 760	635	126
IR.305 + 10 F1	45	60	1 085	2 810	625	145
IR.305 + 10 F3	30	63	1 000	2 610	660	118
SBR 1509 NF	30	60	890	1 620	440	Porous
	45	70	940	1 600	465	60
SBR 1509 + 10 F1	30	66	1 025	1 575	485	63
SBR 1509 + 10 F3	30	66	900	1 540	525	62

Notes: (i) NF refers to no factice
(ii) F1 and F3 refer respectively to a Grade 1 factice, and a Grade 3 factice
(iii) Indentation hardness was determined on the thicker sections of the moulded article. The other physical properties were determined on test pieces cut from the thinner sections.
* low figures indicate high hardness

factice to reduce time to minimum practical cure. Indentation hardness is regarded as the best method of assessing the cure of the mouldings, because of its ability to detect porosity. The samples used in obtaining the extension test data were cut from the thin sides of mouldings from the plant pot mould used in the trials, and the hardness tests were made on the thick base of the mouldings.

The Grade 1 factice had little effect on the physical properties of the vulcanizates and the tendency was for 10 phr of the Grade 3 factice to reduce modulus at 300% slightly, whilst having little effect on tensile strength.

The effect of factice on rate of flow, particularly at injection moulding pressures, was very small, and in relation to the complete cycle time, was insignificant.

The flow properties of these compounds were further studied in a Macklow Smith constant rate extrusion plastimeter. The ram of the machine used for these tests could advance at 1½ in. per min. or at 3 in. per min. The plunger had a cross-sectional area of 0·5 sq. in., the die was $\frac{1}{8}$ in. long and $\frac{1}{16}$ in. diameter. The die pressure readings obtained on the instrument are a measure of compound stiffness. At the ram speed of 1·5 in. per min., the shear rate is approximately 500 sec.$^{-1}$ which is well below injection moulding shear rates; at ram speed 3 in. per min. the shear rate is approximately 1 000 sec.$^{-1}$, which according to Zahler and Murfitt [5] is at the bottom end of the injection moulding range. The results obtained with this plastimeter (Table 8.4) show factice to have a negligible effect on the stiffness of the compound and confirm the results of the machine trials.

Table 8.4 Results on Macklow Smith constant rate extrusion plastimeter

	Pressure on die	
	1·5 in. per min. (shear rate 530 sec.$^{-1}$)	3 in. per min. (shear rate 1 060 sec.$^{-1}$)
Polyisoprene NF	1 020 psi	1 325 psi
Polyisoprene 10 F3	1 160 psi	1 350 psi
SBR 1509 NF	810 psi	850 psi
SBR 1509 10 F3	700 psi	780 psi

Notes:
  (i) NF refers to No Factice; F1 and F3 refer respectively to a Grade 1 factice and a Grade 3 factice
  (ii) Rates of shear in testing the compounds were calculated as follows:

$$^{1} = \frac{4Q}{\pi R^3}$$

Where ^{1} = rate of shear (sec.$^{-1}$)
        Q = volume output (cc./sec.)
        R = radius of die (cm.)

Then for this instrument
        $Q = 0·205$ cc./sec. for 1·5 in./min. ram speed
        $R^3 = 0·0005$ (approx.)

Thus $^{1} = \frac{4 \times 0·2}{\pi \times 0·0005} = 530$ sec.$^{-1}$

PLATE 9

Fig 8.1. Difference in porosity between factice and non-factice compounds.

Fig. 8.2. FRADA high shear plastimeter.

Fig. 8.3. FRADA mincer plastimeter.

PLATE 10

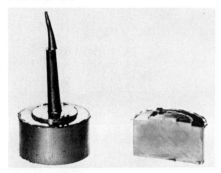

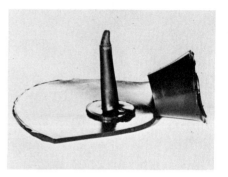

Fig. 9.1. The Peco moulding.
Diameter, 4 cm.; thickness,
2·1 cm. (0·8 in.); weight, 35 g.
of compound SG 1·1.
Courtesy of NRPRA.

Fig. 9.13. The Peco thin sheet
moulding. Diameter, 10 cm.;
thickness, 0·28 cm.; weight
25 g. of compound of SG 1·1.
Courtesy of NRPRA.

Fig. 9.2. The RAPRA beaker moulding, maximum diameter, 7·5 cm.;
height, 12·3 cm.; sidewall thickness, 0·16 cm.; base thickness,
0·95 cm. and weight 80 g. of compound of SG 1·1.
Courtesy of NRPRA.

Table 8.5 Curemeter data at 140°C on 'Ankerwerk compounds'

	Time to 90% PPCL (min.)
Polyisoprene NF	10½
Polyisoprene 10 F3	8
SBR NF	7½
SBR 10 F3	5¾

Notes:
NF refers to No Factice; F3 refers to a Grade 3 factice

Table 8.6 Compounds used for investigating the effect of factice in the Edgwick injection moulding machine

Styrene–Butadiene Rubber (SBR)	*Parts by wt*	*Nitrile Rubber (NBR)*	*Parts by wt*
SBR 1500	100	Breon 1041	100
HAF black	45	HAF black	40
Dutrex R	4·5	DOP	8
Zinc oxide	5	Zinc oxide	4
Stearic acid	2	Stearic acid	1
ZMBT	2·5	TMTMS	0·6
TMTMS	0·6	Sulphur	1·0
Sulphur	3·0	Grade 3 factice	0 and 10
Grade 3 factice	0 and 10		

Natural Rubber (NR)	*Parts by wt*	*Neoprene (CR)*	*Parts by wt*
RSS 1	100	Neoprene WRT	100
HAF black	60	China clay	40
Dutrex R	6	HAF black	30
Zinc oxide	5	Dutrex R	10
CBS	1·0	Zinc oxide	5
TMT	0·15	Stearic acid	3
Stearic acid	2	PBN	1
Sulphur	2	DOTG	0·5
Grade 3 factice	0 and 10	TMTMS	0·5
		Sulphur	3·5
		Light calcined magnesia	4
		Grade 3 factice	0 and 10

Notes:
The compounds (Table 8.6) based respectively on natural rubber, SBR, NBR, and CR, were investigated with and without 10 phr of a Grade 3 factice. The first 3 of these compounds were similar to those used by RAPRA in their preliminary investigations into the injection moulding process. The results obtained on the Edgwick machine were similar to those obtained on the Ankerwerk machine in that factice was observed to reduce the cycle time by as much as 40% (Table 8.7).

71

Table 8.7 Results obtained with a Edgwick 45 SR injection moulding machine

	SBR		NBR		NR		CR	
	NF	F	NF	F	NF	F	NF	F
Mould filling time (sec.)	3·6	2·9	7·6	5·9	> 45	11·0	7	3·5
Mould filling pressure (psi)	11 000	10 000	16 000	15 000	11 000	11 000	11 000	11 000
Temperature of compound before ejection	99	100	103	107	105	112	104	107
Temperature of compound after ejection (°C)	136	138	137	140	138	141	122	132
Minimum practical cure at approx. 180°C (sec.)	35	20	20	15	ND	15	65	50

Notes:
NF refers to no factice
F refers to 10 phr Grade 3 factice
ND = not determined

Unfortunately it is impracticable to determine corresponding cure rates at 200°C in a laboratory curometer because of (i) the inconvenience of preheating the rubber to the temperature at which it emerges from the die of the machine, (ii) heat losses in the curometer at 200°C, (iii) at 200°C the time required to get the test sample on temperature is of the same order as the time required for the cure, which thus appears longer th it is, (iv) that rapid rate of set up at 200°C makes accurate analysis of the curometer data extremely difficult and only time to full cure (as opposed to 90% possible crosslinks, generally regarded as optimum cure) can be assessed with any degree of accuracy. Curometer data at 148°C we obtained (Table 8.5) and show a reduction, due to factice, in cure time fo 90% possible crosslinks; this reduction however, is proportionately not as great as that observed in time to minimum practical cure in the machine at 200°C.

3.3 EDGWICK 45 SR MACHINE TRIALS

The results of the tests in the Ankerwerk machine were of sufficient inter to warrant further investigation. Another range of compounds was therefore studied on an Edgwick 45 SR screw plunger type machine, by the courtesy of the Rubber and Plastics Research Association of Great Britain. The results of this further investigation were described by Flint, Clark and Bibby [6]

The results of this series of investigations are given in Table 8.8; the machine conditions were as follows:

(i) Barrel temperature 90°C
(ii) Mould temperature 180°C

The tendency was for factice to reduce mould filling time to a greater extent than observed on the Ankerwerk machine. The natural rubber compound without factice ejected so slowly that scorch occurred before ejection was complete. The presence of factice, however, removed this difficulty in a striking way by allowing the mould to be filled in an acceptable time. Although reduction in mould-filling time was quite substantial in the case of the other compounds, the total effect on full cycle time was, in these cases, only very small. Had it been possible, with the SBR and NBR compounds, to maintain the same pressure in the factice and the no factice compounds, the differences observed there between factice and no factice would obviously have been greater. Unfortunately, because of the inability of the natural rubber compound without factice

Table 8.8 Processing characteristics of the 'Edgwick test' compounds (laboratory evaluation)

	SBR		NBR		NR		CR	
	NF	F	NF	F	NF	F	NF	F
Mooney viscosity (ML 100°C)	39	34	46	41	31	28	30	20
Mooney scorch (min. at 120°C)	31	25	76	40	13	12	62	42
Minimum practical cure (laboratory, min. at 141°C)	10	6	10	10	3½	3	12½	15
Optimum cure (min. at 141°C)	20	16	15	14	9·3	8·5	75	75
Minimum practical cure (laboratory, sec. at 180°C)	55	35	65	45	27	23	70	55
Optimum cure (laboratory, sec. at 180°C)	110	90	90	60	43	39	460	460
Minimum practical cure in Edgwick machine at approx. 180°C (sec.)	35	20	20	15	ND	15	65	50

NF refers to no factice
F refers to 10 phr Grade 3 factice
ND = not determined

Notes:
(i) Minimum practical cure was estimated from a range of press cures carried out on blanks preheated to simulate ejection heat build-up. The criteria were demoulding (good surface and no distortion), freedom from bloom on standing, tension set, oven ageing and 100% modulus.
(ii) Optimum cure was estimated from conventional laboratory tests applied to a range of press cures.

A comparison of laboratory data giving the optimum cure times of these compounds and the minimum practical cure times, indicates that the effect of factice was greatest at the early stages of cure.

to eject fully, the results are, quantitatively, incomplete; factice has, howe
reduced time to minimum practical cure in all compounds investigated.

Temperature rise on ejection (see Table 8.7) measured by the method
described earlier shows factice to produce an insignificant increase in
temperature rise in all compounds except the polychloroprene. This
confirms the evidence put forward earlier that the temperature rise cause
by factice is not responsible for the reduction in time to minimum
practical cure. This reduction in cure time must be attributed to the
chemical action of the factice on the vulcanizing system.

Table 8.9 Vulcanizate properties at optimum press cure (141°C) of 'Edgwick test' compounds

	SBR		NBR		NR		CR	
	NF	F	NF	F	NF	F	NF	F
Tensile strength (psi)	2 150	2 100	3 400	3 100	3 800	3 500	2 530	2 530
Modulus, 100% (psi)	590	580	224	223	490	350	360	280
Elongation at break (%)	250	215	640	610	450	475	500	500
Hardness (IRHD)	70	70	57	58	75	73	70	65
Optimum cure (min. at 141°C)	20	16	15	14	9·3	8·5	75	75

Notes:
NF refers to no factice   F refers to 10 phr Grade 3 factice

The effect of factice on physical properties can be seen in (Table 8.9).
Factice reduced tensile strength by 10% in the NBR and NR compounds,
and had no effect in the SBR and CR compounds.

8.4 CURING TIME CONSIDERATIONS

An interesting comment on 'minimum practical cure' was published
recently [7] This mentions that to obtain maximum output of moulded
articles, press times are often reduced until the mouldings are just
sufficiently cured to be free from porosity. Residual heat in the moulding
it is suggested, continues the vulcanization after removal of the
moulding from the press. Although the press time necessary to avoid
porosity of course depends on what accelerator is used in the compound,
the point is made that the accelerator most active in reaching optimum
cure is not necessarily the best if porosity is to be avoided when press
time is reduced. It is stated that activity earlier in the curing cycle is the
more important consideration Therefore, it is claimed, Mooney scorch
times are a good criterion, and times to optimum cure are not. In the
laboratory data supporting this intersting discussion, porosity of reduced
cure mouldings is assessed by examination of 'step-wedge' mouldings,
i.e. moulded wedges in which the thickness decreases abruptly by steps.
In a series of accelerators having the same time of optimum cure, activity
earlier in the curing cycle is indicated after 60% cure time by the thicknes

of that part of the wedge showing no porosity. These observations are correlated with Mooney scorch figures. Stated briefly, the claim—an eminently reasonable one—appears to be that to increase productivity by moulding to 'minimum practical cure', the scorch time of the rubber compounds should be reduced to the minimum value that is safe.

Tables 8.10, (a) and (b) The effect of factice on a NR gum stock with a range of accelerator systems

Compound				
	RSS 1	100	100	100
	Zinc oxide	5	5	5
	Stearic acid	1	1	1
	Sulphur	2·0	2·0	2·0
	Accelerator (see Table 8.10 (a))	0·75	0·75	0·75
	Grade 1 factice	—	10	—
	Grade 3 factice	—	—	10

(a)

	a	b	c	d	e	f	g	h	i	j
CBS	0·75	0·6	—	—	—	—	—	—	—	—
TMT	—	0·15	—	—	0·15	—	—	—	—	—
MBTS	—	—	0·75	0·6	0·6	0·6	—	—	—	—
TMTM	—	—	—	0·15	—	—	—	—	—	—
MBT	—	—	—	—	—	—	0·75	0·6	0·6	—
DPG	—	—	—	—	—	0·15	—	0·15	—	—
ZDC	—	—	—	—	—	—	—	—	0·15	0·75

(b)

Accelerator (see Table 8.10 (a))	a	b	c	d	e	f	g	h	i	j*
Mooney set up time at 140°C (min.) NF	21	14	22	12	17	14	8	6	7	5
Time to 90% PPCL at 140°C (min.) NF	26·5	18	33	14	18	19·5	19	13	8	4
Percentage reduction in Mooney set up time due to F1	34	37	31	8	46	41	14	46	54	40
Percentage reduction in Mooney set up time due to F3	42	30	38	4	42	45	19	55	54	40
Percentage reduction in time to 90% PPCL at 141°C F1	42	42	29	Nil	30·5	28	26	23	12·5	Nil
Percentage reduction in time to 90% PPCL at 141°C F3	32	36	24	Nil	39	38	31	Nil	31	Nil

* Mooney test conduct at 120°C

Notes:

(i) NF refers to no factice; F1 & F3 refer respectively to a Grade 1 factice and a Grade 3 factice

(ii) PPCL signifies percentage possible crosslinks.

As already mentioned, factice appears to have most activating effect on the earlier stages of the curing cycle. The test we prefer is a comparison of 'Mooney set up time' determined at as high a temperature as is practicable (e.g. 140°C). The time noted is that at which the Mooney curve starts to rise by 2 points per minute.

As part of our study of factice in the injection moulding process, a range of accelerator systems was investigated in a natural rubber gum stock, with and without 10 phr of a Grade 1 or a Grade 3 factice. Time to 90% possible crosslinks at 140°C and 'Mooney set up time' at 140°C, were determined (Table 8.10).

In the majority of the accelerator systems examined in Table 8.10, factice greatly reduced optimum cure time at 140°C and also reduced 'Mooney set up time' at the same temperature. 'Mooney set up time' measured at curing temperature can be taken as a comparative measure of the tendency for set up (minimum practical cure) to occur early in the period required for a full cure. It is interesting to note that set up time at curing temperature was often proportionately affected more than time for a full cure. This agrees with the evidence, already noted, that factice has its greatest effect at the early stage of vulcanization.

Table 8.11 Compound used in step-wedge moulding

RSS 1	100
Natural whiting	100
Zinc oxide	5
Stearic acid	1
Sulphur	2·0
PBN	1·0
Factice (Grade 3)	0 and 10
TMT	0·05
CBS	0·7

We have used step wedge mouldings to examine the value of factice in controlling porosity when press time is reduced. The mould used by FRADA was a two-piece mould which enabled factice and no factice compounds to be cured side by side, so that moulding conditions in both compounds were identical. The compound shown in Table 8.11 was cured for 60% of no factice optimum cure time at 160°C. The moulds when removed from the press were quenched in water, and the mouldings split down the middle, so that the degree of porosity could be examined. Fig. 8.1 Plate 9 shows a difference in porosity between factice and no factice compounds, and illustrates the effect of factice in reducing porosity of natural rubber compounds at cure temperatures approaching injection moulding temperatures.

8.5 LABORATORY TEST METHODS

Injection moulding machines are expensive items of capital equipment and have to be operated to tight production schedules. It is therefore difficult and uneconomical to use these machines for long term compounding development. Unfortunately, standard laboratory equipment is inadequate for assessing injection moulding compounds. Von Walt and Deskin [8] have shown that marked differences in ML4 values at 100°C may disappear completely at the much higher shear rates associated with injection moulding.

Curometer data do not correspond with cure data obtained from machines trials, for reasons discussed earlier.

Techniques must therefore be developed which are capable of assessing injection moulding compounds in the laboratory. FRADA have developed interesting techniques for this purpose and these were described in the paper by Flint, Clark and Bibby [6]. The most interesting of these techniques was perhaps the FRADA high shear plastimeter, illustrated in Fig. 8.2 Plate 9.

The high shear plastimeter was designed to measure rate of ejection and resulting temperature rise of rubber compounds at shear rates associated with injection moulding. The compound under test is placed in a steel cylinder of cross-sectional area 3·9 sq. cm. terminating in an interchangeable die of varying bore; a die of 3 cm. length and 0·3 cm. diameter is generally used. The cylinder, which is heavily lagged, is heated electrically, and controlled to within ± 1°C by means of a thermostat. The ram is actuated by a hydraulic cylinder connected to a high pressure oil system, operated through adjustable relief valves which enable the extrusion pressure to be set to the desired value. The oil in the pressure system is water cooled, thus making continuous running possible without the danger of possible overheating. Two oil pressure gauges are needed to cover accurately the range from 100 psi to 2 000 psi. The oil line pressure as measured on the gauge, is multiplied by the hydraulic cylinder 11·8 times to give the pressure on the rubber in the barrel, so that the maximum pressure of 2 000 psi on the gauge represents an ejection pressure of 23 600 psi. An electrically operated timing device reading to 0·1 of a second is built into the instrument and registers the time taken to eject a constant volume (41 cc.) of the compound under test.

Measurement of rate of flow is a relatively simple matter provided standard procedures are observed. Measurement of temperature rise on ejection, however, presented certain difficulties. Temperature rise was originally measured by stabbing with a needle thermocouple. It was found, however, that to get reasonably accurate and reproducible results, refinements inconvenient for routine work were essential. A simpler method, which was accurate and capable of being used for routine experiments, was therefore devised.

This was a calorimetric method which utilized an ordinary domestic vacuum flask of known water equivalence. The compound was ejected directly into the flask containing 350 cc. of water at a known temperature. The final temperature of the water was measured and from a knowledge of the specific heat of the compound, the temperature of the extrudate could be calculated. As this was based on fundamental physics the results were accepted as accurate. Table 8.12 shows differences between temperature rise as measured by stab temperature techniques and by calorimetry.

Table 8.12 Temperature rise caused by ejection at high shear rate

	SBR		NBR		NR		CR	
	NF	F	NF	F	NF	F	NF	F
High shear plastimeter (11 800 psi barrel pressure, die 3 cm. long 3 mm. diam.)								
Rise measured by calorimetry (°C)	26	22	38	33	35	30	35	25
Rise measured by 'stab temperature'(°C)	10	8	20	19	11	8	17	12
Edgwick injection moulding machine (for ejection pressures see Table VII)								
Rise measured by 'stab temperature'(°C)	37	38	34	33	33	29	18	25
Ejection time (sec.)								
Edgwick machine (mould filling time)	3·6	2·9	7·6	5·9	>45	11·0	7·0	3·5
H.S. plastimeter (3 mm. bore die, 6·8 cm. long and barrel pressure 11 800 psi)	8·7	7·5	100	33	8·8	7·6	9·9	4·9

Notes:
NF refers to no factice   F refers to 10 phr Grade 3 factice

(i) 'Stab temperature' is determined by catching the extruded compound in a rubber cup, squeezing together rapidly by hand and at once noting the temperature with a calibrated needle thermocouple.

(ii) The temperature 'by calorimetry' was calculated from the rise in temperature observed when a known volume of compound was ejected into a known volume of water in a domestic vacuum flask. The specific heat of the compound had previously been determined.

(iii) For the comparison of time of ejection by means of the high shear plastimeter, a 3 mm. bore die of unusual length (6·8 cm.) was selected to bring the conditions more into line with those obtaining in the nozzle and. mould of the Edgwick machine.

The rates of extrusion and heat build-up on ejection cannot be directly compared with the results obtained from the injection moulding machine because (i) the action of the FRADA high shear plastimeter more resembles the action of a ram type machine than a screw plunger type machine, and (ii) the die used in the laboratory machine could not be equated to the runners in the mould of the Edgwick machine. The results are, however, comparable as between factice and no factice under the same experimental conditions.

During the machine trials the tendency was for factice to cause an increase in temperature rise on ejection. In the case of the high shear plastimeter, however, factice reduced temperature rise on ejection. This difference was attributed to the action of the screw in the screw plunger machine.

Table 8.13 Results of 'mincer plastimeter' test stimulating screw action of injection moulding machine

	SBR		NBR		NR		CR	
	NF	F	NF	F	NF	F	NF	F
High shear temperature rise prior to mincer plastimeter treatment (°C)	26	22	38	33	35	30	35	25
Final temperature reached by end of mincer treatment (°C)	101	100	109	104	89	86	92	90
Test data after mincer treatment for 2½ mins								
High shear ejection time (sec.)	2·9	2·5	2·9	2·6	3·6	2·9	2·0	1·8
High shear temperature rise (°C)	26	26	35	33	28	28	30	22

Notes:
NF refers to no factice   F refers to 10 phr Grade 3 factice

(i) High shear ejection time is time in sec. to extrude 41 cc. of compound at 11 800 psi through a die 3 cm. long and 3 mm. diameter. The initial temperature of the compound was 90°C.

(ii) High shear temperature rise was measured by ejecting into the calorimeter when H.S. ejection time was determined and is the difference between final extrudate temperature and the initial temperature of 90°C.

(iii) After the period of treatment in the 'mincer plastimeter', the compound was at once transferred to the high shear plastimeter with as little loss of heat as possible and tested a few minutes later when the temperature of the compound had stabilized at 90°C.

In order to determine the effect of this screw on compounds to be ejected at high shear, an instrument termed the FRADA Mincer plastimeter was employed to simulate the action of the screw. This instrument (Fig. 8.3 Plate 9) was made from an electrically operated meat mincing machine which was modified by redesigning the scroll and completely blocking the barrel end, so that instead of being extruded by the action of the scroll, the rubber compound was forced to circulate. To simulate injection moulding conditions, where access of air is restricted, nitrogen gas was passed into the barrel. The machine is well instrumented, giving temperature rise, torque on the scroll, revolution count, and power input. The temperature of the heavily lagged barrel is electrically controlled. The instrument was designed particularly for measuring the running temperature and processing scorch of rubber compounds. High shear plastimeter results given in Table 8.13 show that prior action of the screw tends to reduce the cooling action of the factice on ejection temperature rise.

8.6 SUMMARY

In summary, the experimental data presented in the above paper, lead to the conclusion that factice is a useful ingredient in rubber compounds to be processed by injection moulding. The effect of factice is to reduce the time required for minimum practical cure, because factice promotes set up early in the curing cycle. Reductions in minimum practical cure time as great as 50% may be obtained, and it would appear that the magnitude of this effect is dependent on the accelerator system used, as well as on the polymer on which the compound is based, and on other details of the formulation. There is evidence that factice controls the development of porosity in minimum practical cures.

There is also a tendency for an addition of factice to reduce mould-filling time, although usually not in the same proportion as the effect on minimum practical cure. However, mould-filling time is so small a part of total cycle time that the effect of factice is only of minor importance: the major effect is in producing substantial reductions in time for minimum practical cure. This is obtained by using only a relatively small amount of factice in the compound (e.g. 10 phr) and is achieved with negligible effects on the physical properties of the moulding.

ACKNOWLEDGEMENTS

The authors thank the Council of the Factice Research and Development Association for permission to publish this paper.

REFERENCES

1 Flint, C. Falconer, Factice compensated acceleration *Rubb. J. and Intern. Plast.* May 1959.
2 Flint, Wardle and Clark, Pattern of properties *Proc. of the 4th Rubber Tech. Conf.* London 1962.
3 Morris and Izod, *Rubb. and Plast. Age* **46**, 2, 167–172.
4 Quirk and Held, *Rubber Age* **99**, 10, 84–89.
5 Zahler and Murfitt, *Brit. Plast.* Dec. 1963, 698.
6 C. F. Flint, A. H. Clark and N. H. Bibby, Use of factice in the injection moulding process applied to rubbers. Proc. International Rubber Conference 1967, Maclaren and Sons Ltd 1968.
7 ICI Ltd, Technical Information R136.
8 Von Walt and Deskin, Du Pont de Nemours and Co., Elastomer Chemical Department Contribution No. 229.

DISCUSSION

W. S. PENN *(Borough Polytechnic)*: Why does factice speed up cure?

ANSWER: There is some evidence that the activating effect of factice is temperature dependent, i.e. the higher the temperature, the greater the effect of factice on activation. It is thought that the polysulphide linkages in the factice molecule tend to rupture at elevated temperature producing active sites within the factice molecule, or active free sulphur.

Q: Can you explain the method of melt temperature determination by the vacuum flask method, particularly the difficulty of getting the melt into the flask?

A: The calorimetric method was only used on rubber compounds in connection with the laboratory instrument as there the technique of getting the rubber into the flask presented no difficulties. The instrument is vertical and the flask containing the water at known temperature can be placed directly beneath the nozzle, and the rubber injected directly into the flask.

Although a method of this type is necessary for laboratory determinations where only 41 cc. of compound is being ejected, and heat losses can be great, it may not be essential for a 4 oz. injection moulding machine where the colume of rubber ejected is far greater, and heat losses from the centre of the mass of consolidated rubber obviously less.

MR BOOTH *(Esso Research SA)*: commented on the sensitivity of temperature determinations. In using a thermocouple (the so-called stab method) one needs (a) more than 30 gm. melt to avoid considerable heat loss to atmosphere (b) a sensitive thin needle (c) more than one stab The result will probably be accurate to within about 3–5 degress of the calorimetric physical method.

CHAPTER **9**

Injection moulding of natural rubber

M. A. WHEELANS, Ph. D., B.Sc., A.N.C.R.T., A.I.R.I.
The Natural Rubber Producers' Research Association

9.1 SUMMARY

IT is the purpose of this chapter to show how a wide variety of natural rubber (NR) compounds may be successfully injection moulded in short cure times using six different moulding machines in conjunction with a larger number of different moulds. By appropriate attention to machine variables quite conventional NR compounds may be moulded without resorting to special compounding. In particular cases, however, as for example in the moulding of thick sections adjacent to thin ones, the recently developed 'efficient vulcanization' (EV) curing systems may be recommended.

The function of an injection moulding machine is to plasticize a rubber compound and to inject it into a closed mould in which it is cured. Machine settings are capable of wide variation and the rubber manufacturer will aim at the best combination of these to give the maximum rate of production. One of the objects of the work to be described was to understand the effects of the machine controls so that the maximum possible use could be made of the injection moulding machine/natural rubber compound combination. It is now clear that the highest rates of cure may be obtained by achieving the best of a compromise between injecting rubber at the highest possible temperature while maintaining freedom from scorch.

9.2 INTRODUCTION AND REVIEW OF PROCESS

Perhaps the most important feature of injection moulding is the large element of automation that it allows the rubber industry to introduce into its moulding operations. A number of authors [1-14] have discussed the advantages of this process over conventional compression moulding—the industry's most popular processing method—and concluded that:

1 Preparation of rubber for injection moulding is simpler because the cutting, shaping and weighing of blanks and the manual handling of moulds required for conventional compression systems can be eliminated.

2 Cure times can be cut to one-tenth [1] or one-twentieth [2] by using high curing temperatures and injection temperatures

Photographic illustrations for this chapter Figs 9.1, 9.2, 9.13, 9.22, 9.24, 9.25, 9.26, 9.27, 9.28, 9.29, 9.30, 9.31, 9.32, 9.33, 9.34, will be found on Plates 10-15

almost as high as that for curing.

3 Quality of products can be improved in uniformity of dimensions by injection into a closed mould, and in uniformity of physical properties by the regularity of the automated cure cycle.

4 Moulds with self-trim grooves can be used to give virtually flashless mouldings, thereby eliminating or reducing trimming and inspection costs.

5 Material costs can be reduced by eliminating flash and by reducing the number of rejects [3, 4] due to oversize or flow deformities.

In addition to advantages of better quality control and more efficient use of labour it was reported recently [13] that greater versatility of style and colour was possible by injection moulding of footwear. It added that better working conditions followed from injection moulding and that a new attitude to work resulted if an operative could take a pride in working a machine to make a whole shoe instead of being responsible for only a small part of it.

These advantages have to be set against the high cost of injection equipment and some of these claims have been criticized as oversimplifications. For example, Gregory [12] points out that the cost of providing continuous strip feed or comminuted rubber has to be considered against rubber preparation costs for compression moulding. He also notes that, to achieve reduced manual handling, mould complexity increases—possibly requiring duplicate plates, removable and interchangeable cones and perhaps mechanically operated 'shuffling' plates.

Even so, it is now accepted by the rubber industry that injection moulding can provide valuable processing savings [3, 4, 8, 12, 13] and as a consequence, it is now an established process—especially where long production runs of automotive components are involved.

In a particularly frank lecture to the RAPRA symposium on 'Higher Temperatures in the Processing of Rubber', Dr A. P. C. Cumming [14], then director Polymer Engineering Division, the Dunlop Company Ltd, stated that injection moulding at present shows about 25% savings over conventional moulding and that further substantial savings could be expected from improved mould design to produce flash-free, trimless mouldings.

It is understandable that the process is becoming more widespread and consequently the number of injection moulding machine manufacturers is rapidly increasing [15, 16].

A brief introduction to the history [17] and nature of the process mentions that in 1943 a patent for a machine capable of injection moulding natural and synthetic rubber was applied for by W. P. Cousino, Chrysler Corporation [18], and that in 1946 R. T. Vanderbilt published a handbook [19] on injection moulding of rubbers with the Chrysler (Monroe) equipment. However, in spite of developments [20, 21] providing greater injection power and improved equipment designs [22], the process was slow to gain wide acceptance—even though some especially complicated mouldings requiring numerous preshaped moulding blanks (e.g. milking inflators) were satisfactorily injection moulded.

In the early 1960s a reawakening of interest in injection moulding of rubber was initiated by a new generation of injection units [15, 16]. Developments consisted of modifications to simple plunger machines (e.g. the Seidl rubber machine, first exhibited at the 1961 Pirmasens Fair), and, in particular, to reciprocating screw machines. These developments were aimed at improving process control to avoid scorch and loss of material through seals.

The great advantage of modern reciprocating screw machines stems from their ability to generate heat, which can be regulated by screw speed, back pressure and by thermostatically controlled coolant fluid (water, glycol, or oil). The rubber is heated and plasticized as it progresses along a retractable screw. It then masses at the front of the backward moving screw until a switch is tripped, signalling injection by the forward plunger action of the screw. The final increment in heat is provided by injection through a narrow orifice into the mould.

Plasticization by screw is quicker, more controllable and gives a more homogeneous distribution of heat and viscosity than that given by a simple plunger system. It is argued, however, that engineering of the simple plunger is cheaper [23]

Of the 47 rubber injection machines available in October 1963, 85% were screw machines and 15% were plunger machines. Only one machine out of the 43 announced between October 1963 and December 1964 was a plunger machine [16]. It is therefore appropriate that most attention should be devoted to the reciprocating screw type of machine.

Both machine variables and rubber compound variables have a marked effect on rates of cure, rates of scorch and general injection moulding performance. The effects of machine variables were examined first and, when convenient conditions had been established, rubber compounds were examined systematically.

The injection moulding machine variables may be divided into two groups as shown by Harrison and Seklecki [24]:

Group 1   Those which act to control the heating and plasticity of the rubber during the first or preparatory stage when the volume to be moulded is being charged into the front of the injection chamber.

Group 2   Those which influence the second stage of injection unit operation, that of mould filling.

Primary consideration was given to group 1 variables because they are independent of mould characteristics and, therefore, give flexibility to process adjustment without mould modification.

In group 1 the following variables were investigated:

(a)   Screw speed,
(b)   barrel temperature,
(c)   screw back pressure (the adjustable hydraulic pressure which acts to pressurize the plasticized 'melt' in the injection chamber during screw rotation).

In group 2 the variables were as follows:

(d)   Nozzle orifice diameter,

(e)   injection pressure,

(f)   rate of injection.

To a large degree these variables are interdependent.

Other variables examined included nozzle temperature and mould temperature.

9.3 EXPERIMENTAL DETAILS

Two reciprocating screw machines were examined in detail. The Peco 21TS [24] machine and the Peco moulding described in Fig. 9.1 Plate 10, were used for the examination of machine variables as outlined above and the Daniels Edgwick 45 SR [25] machine was used for examination of a more limited number of variables reported by Wheelans [26, 27, 28].

The bulk of the work described in connection with variation of rubber compounds was carried out on the Daniels Edgwick 45-SR [25] injection moulding machine using a particularly searching beaker shaped mould (Fig. 9.2 Plate 10) designed by RAPRA. The mould described by Izod and Morris [29], was designed to accentuate flow difficulties and at the same time to show differences in state of cure of different thickness mouldings.

Other screw machines which have been very successfully used for research and development trials with NRPRA compounds are the 4-station Desma 905 [30] machine, the Stübbe S150/235 [31] and SKM 75/80 [31] machines and the Ankerwerk [9, 32] V17-65 machine.

The Farrel-Bridge R-60-350 [11] simple hydraulic plunger machine and the Seidl SPA 1 BX-A piston machine [33] with novel ball bearing packed plasticization cylinders have also been satisfactorily used in trials.

Injection time was measured by observation of the forward movement of the ram. Injection pressure was measured by a pressure gauge in the line from the pump to the ram (1 200 psi line pressure on the Daniels 45 SR machine is equivalent to 16 400 psi material pressure).

Injection temperatures were measured by a thermocouple probe inserted into a mass of rubber injected under pressure through the nozzle into an insulated cup at 30 sec. intervals. The temperature of stock emerging from the barrel was measured by the same probe immediately the nozzle had been removed. This is referred to as 'extrudate temperature'.

Samples were taken at preset moulding conditions until several successive thermocouple readings, taken on a cut section of the sample, showed that consistent conditions were established.

At each selected setting several samples were moulded at progressively increasing cure times. These samples were tested for hardness or micro-hardness on the top, bottom and in a central section and the results collated on a hardness *vs* cure time chart.

Tensile test pieces were cut from the side walls, perpendicular to the base of the beaker. Trouser tear strips were also taken from the side wall but were cut parallel to the base. Hardness and Dunlop resilience testing was carried out on the base of the beaker.

The most studied base mix (see Mix 1, Table 9.6) contained NR, 100 parts and SRF black, 50 phr, because it was felt that this type of mix would have utility in the manufacture of engine mountings, engineering rubbers, shock absorbers and, or other automotive parts. The most investigated cure system consisted of sulphur 2·5 phr, and CBS (Santocure) 0·5 phr, and this was chosen for its high level of safety and general applicability. Variations in filler type and in curing systems are described.

9.4 EFFECTS OF MACHINE VARIABLES ON INJECTION MOULDING BEHAVIOUR

9.4.1 GROUP (1) VARIABLES AFFECTING PLASTICIZATION
(Peco 21TS machine [24])
1a   Screw speed

The effect of screw speed on injection temperature was not very noticeable at the beginning of experimentation because at slow speeds cure times were relatively long. During long (2–3 min.) cures the cooling effect of the relatively low barrel temperature dominated the rubber temperature situation in the chamber in front of the ram, especially when screwing back occurred relatively soon after injection. When a screw delay timing device [24] was operated and screwing back to fill the injection chamber was arranged to take place at a fixed time *before* injection, the injection temperature increased with changes in screw speed as shown in Fig. 9.3.

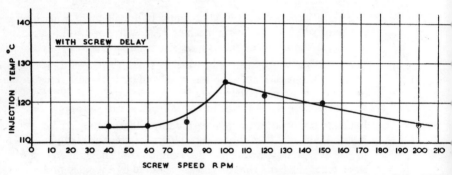

Fig. 9.3. Variation of injection temperature with screw speed. Peco 21TS machine [24] and ⁷⁄₈ in. thick mould operating screw delay procedure, mould temperature 175°C, barrel temperature 90°C; injection pressure (line), 1 000 psi; screw drive pressure, 950 psi; screw back pressure 150 psi; nozzle diameter 0·125 in. Mix 1, Table 9.6.

A diagram of cure time (assessed by the hardness of top, bottom and centre of the moulding) *versus* screw speed shown in Fig. 9.4 indicates how closely cure time follows injection temperature and screw speed. Figures 9.3 and 9.4 show that the optimum screw speed is about 100 rpm under the prevailing back pressure and barrel temperature conditions. Speeds greater than 100 rpm caused poor consolidation of the feed and subsequent air trapping.

PLATE 11

Fig. 9.22. The RAPRA frustum cone mould. Diameters: minimum, 4·8 cm.; maximum, 5·2 cm.; thickness, 3·7 cm. (1·5 in.); weight 90 g. of compound SG 1·1.
Courtesy of NRPRA.

Fig. 9.24. The David Bridge wheelbarrow wheel moulding. Solid diameter of tyre cross-section, 1·9 in.; diameter of wheel, 8 in.; weight, 1 150 g. of compound SG 1·1.
Courtesy of NRPRA.

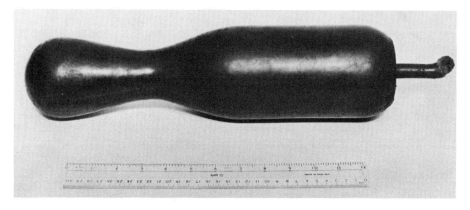

Fig. 9.25. The Desma bowling pin (diameter, 9·8 cm.; 3·85 in.; length, 40 cm., 15·7 in.; weight, 2·43 kg. of rubber, SG 1·136).
Courtesy of NRPRA.

PLATE 12

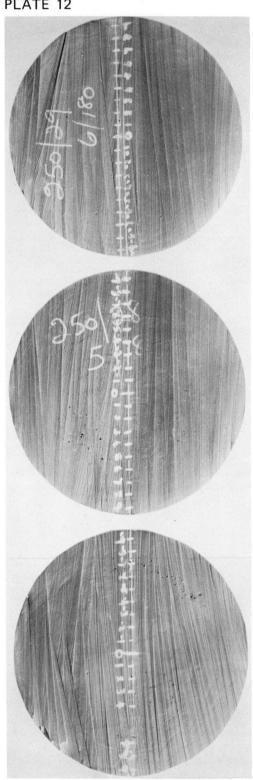

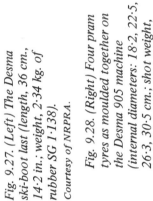

Fig. 9.26. Samples of the Desma bowling pin cut into discs to illustrate the state of cure of NR/HAF black mixes after 4, 5 and 6 min. cure at 180°C as shown in Table 9.16. Four and five min. cures are only slightly porous.

Fig. 9.27. (Left) The Desma ski-boot last (length, 36 cm., 14·2 in.; weight, 2·34 kg. of rubber SG 1·138).
Courtesy of NRPRA.

Fig. 9.28. (Right) Four pram tyres as moulded together on the Desma 905 machine (internal diameters: 18·2, 22·5, 26·3, 30·5 cm.; shot weight, 1·605 kg. of rubber, SG 1·138).

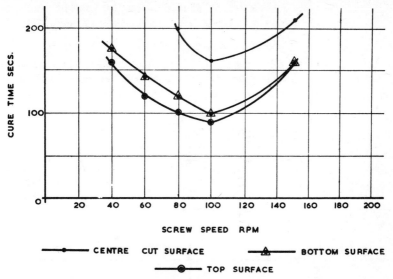

Fig. 9.4. Variation of cure time of surfaces and centre of ⅞ in. thick
mould with screw speed. Peco 21TS machine as Fig. 9.3.

Key   ● Centre of cut moulding
      △ Bottom surface
      ○ Top surface (sprue face)

1b Barrel temperature

Injection temperature increases with elevation of barrel temperature as
shown in Fig. 9.5. When there is no screw delay in operation, barrel
temperatures above 105°C cause scorch. Figure 9.6 shows that higher
barrel temperatures may be used when the screw delay procedure is
employed to maintain the rubber at its pre-injection temperature for
the minimum period and that injection temperatures over 140°C can be
attained without fear of scorch.

1c Back pressure

Injection temperature increases with back pressure as shown in Fig. 9.7,
but at 80 rpm there was no advantage in using a back pressure greater
than 200 psi. Without the screw delay procedure there was little
sensitivity to back pressure changes.

9.4.2 GROUP (2) VARIABLES AFFECTING MOULD FILLING

2d Nozzle orifice diameter

The dependence of injection temperatures and injection times on
nozzle diameter are clearly shown in Fig. 9.8. Injection temperature
increases as the nozzle diameter is reduced to 0·075 in. At diameters
below this, injection times and risk of scorch during mould filling
increase markedly with no advantage in increased injection temperature.
Under small diameter conditions such as these, injection pressure is
fully realized and dominates injection temperature conditions.
Experience based on long injection times causing scorch showed that
the most suitable nozzle diameter for the particular mould used

(Fig. 9.1) was 0·1–0·125 in. although this was wider than the diameter desirable for optimum heat ·build-up through the nozzle.

2e   Injection pressure

Under relatively wide nozzle conditions (0·125 in.) an increase in injection pressure affords a steady rise in injection temperature and a useful reduction in injection time as shown in Fig. 9.9.

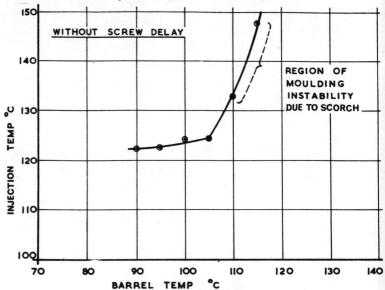

Fig. 9.5. The effect of barrel temperature on injection temperature without screw delay. Peco 21TS machine 40 rpm; other details as Fig. 9.3.

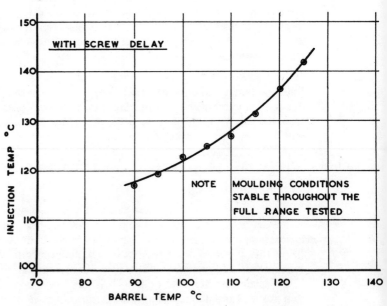

Fig. 9.6. The effect of barrel temperature on injection temperature with screw delay. Peco 21TS machine, other details as Fig. 9.5.

2f Rate of injection

The dependence of injection temperature and cure time on injection ram speed is shown in Figs 9.10 and 9.11. Higher ram speeds give higher injection temperatures and quicker curing, but at the highest ram speeds brought about by reinforcing the pump capacity of the

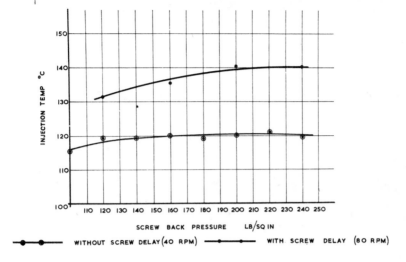

Fig. 9.7. Variation of injection temperature with screw back pressure. Peco 21TS machine with screw delay, barrel temperature 105°C, screw speed, 80 rpm, other details as Figs. 9.3 and 9.4.

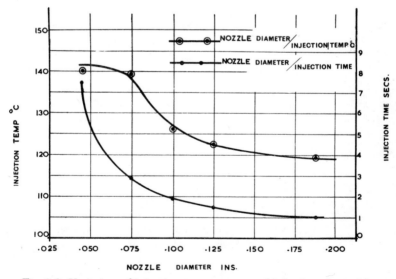

Fig. 9.8. Variation of injection temperature and injection time with nozzle diameter. Peco 21TS machine, barrel temperature 105°C, screw speed, 40 rpm; injection pressure, 1 000 psi (line); other details as Fig. 9.3.

Key ○ *Injection temperature*
     ● *Injection time*

Peco 21TS machine, there was a falling off of the advantages to be obtained.

The work described so far in this section shows that a natural rubber compound can be moulded satisfactorily under a wide variety of conditions in a versatile machine. It demonstrates how machine variables can be used at their optima to give the advantages of rapid curing by obtaining high injection temperatures and quick mould filling times. Figure 9.11 shows that when the injection temperature is about 140–145°C the ⅞ in. thick moulding can be cured in less than 60 sec.

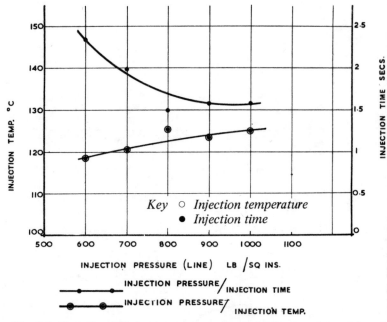

Fig. 9.9. Variation of injection temperature and injection time with injection pressure. Peco 21TS machine, barrel temperature 105°C, nozzle diameter, 0·125 in., screw speed 40 rpm, other details as Fig. 9.3.

9.4.3 VARIABLES ON THE DANIELS 45 SR

While still considering the effects of machine variables on injection moulding behaviour it is worth reporting in some detail the results of a shorter survey carried out with the Daniels 45 SR machine. This was necessary before an examination of compounding variables could be made with this machine and is also of considerable interest because of the nature of the mould used. The RAPRA beaker mould (Fig. 9.2 Plate 10) presents a different situation because its thin walls offer more resistance to the mould-filling pressures than the simple Peco mould and avoidance of scorch by quick injection is therefore much more of a problem during mould filling.

Screw speed

If screw speeds are too high they tend to cause scorch in the barrel. Izod [34] found that 100 rpm was too fast with the Daniels Edgwick

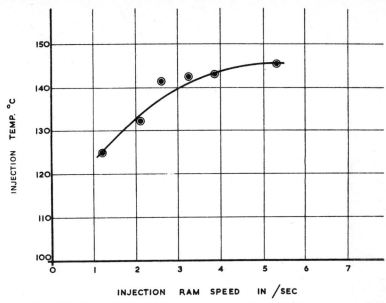

Fig. 9.10. The effect of injection ram speed on injection temperature. Peco 21TS machine, barrel temperature 105°C, screw speed 40 rpm, other details as Fig. 9.3. Ram speed varied by increase in volumetric delivery.

45-SR machine and for the present work the screw speed was fixed at 55 rpm by the internal gear system.

EFFECT OF INJECTION PRESSURE ON INJECTION TIME AND TEMPERATURE

The most advantageous pressure to use from the point of view of obtaining maximum heat build up through the nozzle is the maximum available. The maximum pressure may be safely used provided it does not lead to scorch.

Reduction of injection pressure has the effect of increasing the injection time, and decreasing the injection temperature. Also it may intensify the effect of scorch in a moulding.

Table 9.1 shows that when the maximum available pressure was reduced by 33% the injection time was more than doubled. This finding was consistent with observations by Izod and Morris [29] who found with this machine and mould that quick injection time was critically dependent on injection pressure below a certain value, (10 000 psi, material pressure) and it was concluded that, as far as the present

Table 9.1 Effect of injection pressure on injection time and temperature

Injection pressure psi	15 000	10 000
Injection time sec.	5·6	12·0
Injection temperature °C	127	109

Mix, 1 (Table 9.6); Mooney viscosity, ML3, 120°C, 51; Mooney scorch 120°C, T5, 29 min.; Nozzle diameter, $\frac{1}{8}$ in.; barrel temperature, 90°C; mould temperature, 180°C.

91

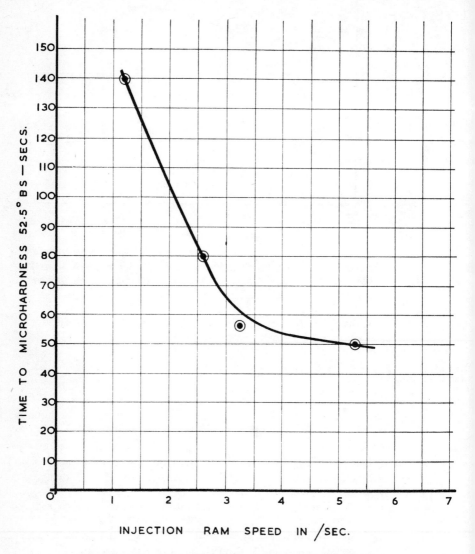

Fig. 9.11. *The effect of injection ram speed on cure time. Details as Fig. 9.10.*

work was concerned, there was no advantage in operating the machine at less than the maximum available pressure.

Only very soft stocks which tend to fill a mould before air is fully expelled demand a reduction of injection pressure.

EFFECT OF NOZZLE DIAMETER ON INJECTION TIME

It is important to fill a mould in a short space of time especially when a rubber compound may have to flow through a long narrow path as in the beaker mould. Changes in nozzle diameter provide a very sensitive means of obtaining short injection times. Table 9.2 shows that, under the conditions stated, the change of nozzle size from $\frac{1}{8}$-$\frac{5}{32}$ in. was the

Table 9.2 Effect of nozzle diameter on injection time

Nozzle diameter in.	$\frac{1}{8}$	$\frac{5}{32}$
Injection time sec.	8·8	3·8
Injection temperature °C	120	118
Appearance of mouldings	scorched	satisfactory

Mix, 1 (Table 9.6); Mooney viscosity, ML3, 120°C, 59; Mooney scorch 120°C, T5, 26 min.; barrel temperature, 90°C; mould temperature, 180°C; injection pressure, 16 000 psi.

only alteration required to make possible the production of a good moulding, free from scorch in the side wall. It also illustrates that the scorch time of the compound was between 3·8 and 8·8 sec. during its flow through the mould.

NOZZLE TEMPERATURE

Scorch may occur in a nozzle in close contact with a hot mould, but when it was found that water cooling caused a cool plug and led to slow injection, no further control of nozzle temperature was attempted.

EFFECT OF MOULD TEMPERATURE ON MOULDING PROPERTIES

Mould temperatures of 180° and 200°C were quite satisfactory with barrel temperatures of 90°C which gave injection temperatures of 120–130°C. When the barrel temperature was elevated to 120–125°C and injection temperatures increased to 150–160°C the most suitable mould temperature was found to be 180°C. Scorch on the mould side walls occurred at higher mould temperatures.

EFFECT OF BARREL TEMPERATURE ON INJECTION TEMPERATURE, INJECTION TIME AND CURE TIME

Elevation of the barrel temperature has the triple advantages of increasing injection temperature and decreasing both injection time and cure time (Figs 9.12a, b and c). The effect on cure time is also shown in Table 9.3.

Table 9.3 Effect of barrel temperature on cure time

Barrel temperature °C	90	100	125
Time to cure 1·6 mm. thick side wall, sec.	45	30	30
Time to cure 1 cm. thick base, sec.	180	90	45

Mix, 17 (Table 9.8); Mould temperature, 180°C; injection pressure, 16 000 psi; nozzle diameter, $\frac{5}{32}$ in.

If other machine settings are approximately correct it is then possible to raise the barrel temperature in order to get the fastest possible cure times. If there are signs of scorch in the barrel or nozzle, the barrel

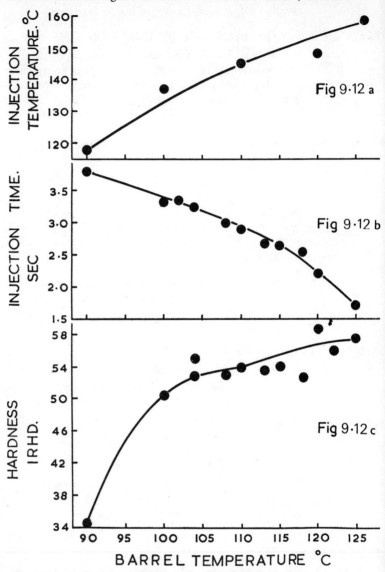

Fig. 9.12. As the barrel temperature rises the injection temperature rises (Fig. 9.12a); the injection time decreases (Fig. 9.12b); the hardness of the 1 cm. thick base cured for 60 sec. at 180°C increases (Fig. 9.12c).

Mix	17
Mooney viscosity	55
Mooney scorch	28
Daniels 45 SR machine	
Nozzle diameter in.	$\frac{5}{32}$
Injection pressure (max.) psi	16 000
Mould temperature °C	180

temperature may be slightly reduced to give adequate safety.
If mouldings become scorched, reduction of mould temperature is
possible. If injection times become so short that air trapping is
apparent, with soft stocks, mould venting or vacuum moulding may
be necessary. Alternatively, the nozzle diameter may be reduced to
increase mould-filling time and allow time for air to be expelled from
the mould. This latter step should also enhance the injection
temperature.

9.5 EFFECT OF RUBBER COMPOUND VARIABLES ON INJECTION MOULDING BEHAVIOUR

The properties of a finished product govern mix composition.
However, it is important to know general effects of changes in
compounding and whether they may be used to avoid difficulties
such as those due to scorch in a moulding.

Scorch may occur during passage of the compound through the screw
(or in the heating chamber of a simple ram machine), the nozzle, the
runners, the gates or the mould itself. Scorch in the barrel may be
observed by abnormally long injection times or, in serious cases
by injection of crumbed material. Scorch in the nozzle may be
sufficiently serious to prohibit any injection at all. Scorch in the
runners may appear as crumb in the moulding. Mild scorch in the
moulding appears as rippling on the moulding walls.

9.5.1 EFFECT OF MOONEY VISCOSITY ON INJECTION TIME

At the same ram pressure and nozzle diameter a mix with a high
Mooney viscosity takes longer to fill a mould than a similar mix with
a lower viscosity (see Table 9.4). If the mould filling time is long a
compound may scorch while the mould is filling. Where scorch due to
this cause is a problem, reduction of Mooney viscosity by mastication,
peptization or oil extension may provide a useful solution by reducing
the injection time, as long as a compound with reduced viscosity will
meet specifications for physical properties of the product and reduced
injection temperature does not lead to a serious increase in cure time.

Table 9.4 Effect of Mooney viscosity on injection time

Mix	1a	1b	1c	2
Mooney viscosity, ML3, 120°C*	59	51	36	38
Mooney scorch, 120°C, T5 min.	26	29	23	34
Injection time, sec.	8·8	5·6	3·7	4·3
Injection temperature °C	120	127	109	110
Appearance of mouldings	scorched	slightly scorched	good	good

Nozzle diameter, $\frac{1}{8}$ in. (chosen to exaggerate scorch); barrel temperature,
90°C; mould temperature, 180°C; injection pressure, 16 000 psi.

* Differences in viscosity of mixes 1a, 1b, 1c were contrived by different degrees
  of mastication. Mix 2 consisted of the same base mix as mix 1 (Table 9.6)
  with the exception that 20% of the volume of NR and 5 phr Dutrex R were
  replaced by 21·6 parts by weight of Sundex 8125 oil.

9.5.2. EFFECT OF INCREASING THE SCORCH SAFETY OF A COMPOUND ON CURE TIME AND MOULDING APPEARANCE

Scorch in the side walls of the moulding may also be avoided by compounding variations to increase the Mooney scorch time. However, if the barrel temperature is fixed as Table 9.5 shows, scorch may only be avoided at the cost of an undesirable increase in the cure time.

Table 9.5 Effect of increasing the scorch safety of a compound on cure time and mould appearance

Mix*	1	3	4	5
Sulphur	2·5	2·5	0·33	–
CBS	0·5	0·5	5·0	–
Retarder	–	1	–	–
MDB (Morfax)	–	–	–	3·56
Mooney scorch, 120°C, T5 min.	26	36	39	45
Mooney viscosity, ML3, 120°C	59	60	55	63
Injection time, sec.	8·8	8·7	7·8	9·2
Injection temperature, °C	120	116	120	119
Time to optimum cure of side walls, sec.	30	60	90	180
Time to optimum cure of 1 cm. thick base, sec.	180	240	240	300
Appearance of mouldings	scorched	slightly scorched	good	good

Nozzle diameter, ⅛ in. (chosen to exaggerate scorch); barrel temperature, 90°C; mould temperature, 180°C; injection pressure, 16 000 psi.

* Other mix ingredients were as mix 1 (Table 9.6).

If an increase in safety also permits an increase in barrel temperature and injection temperature, cure time may be advantageously reduced as shown later in Table 9.12.

9.5.3 EFFECTS OF VARYING CURE SYSTEM AND FILLERS

Long injection times are undesirable and with the larger nozzle (⁵/₃₂ in. diam.) it is possible to mould 'scorchier' and faster curing compounds than those already mentioned. Examples given in Table 9.6 show that a compound containing a S/CBS curing system boosted with TMTD, a TMTD 'sulphurless' compound, and compounds containing high loadings of black fillers such as SRF, MPC and FEF blacks can be moulded satisfactorily. This table also includes a compound containing 50 phr HAF black which is fast curing and yet safe enough in a curing system to be moulded using ⅛ in. diameter nozzle.

Table 9.6 shows that, under the conditions employed, compounds having Mooney scorch times of 9 min. at 120°C were moulded satisfactorily but a compound (not tabulated) containing 150 phr SRF black and having a scorch time of only 6 min. was too 'scorchy' during injection through the ⁵/₃₂ in. nozzle to give good mouldings. There was no problem in moulding black mixes with Mooney viscosities

(ML3, 120°C) as high as 87, and aluminium silicate mixes as high as 110 (Table 9.9).

Table 9.6 compares properties of injection moulded vulcanizates with press cured vulcanizates and also shows that with conventional curing systems some reversion of the side walls takes place before the base of the beaker is fully cured.

9.5.4 EV CURING SYSTEMS

A conventional curing system is quite satisfactory for short injection moulding cures of uniformly thick sections. However, greater reversion resistance is desirable for thin sections adjacent to thicker ones (especially when there is a large disparity between injection and mould temperatures) and this may be obtained by compounding with efficient vulcanization (EV) curing systems [35, 36, 37].

EV systems give vulcanizates which are markedly superior to those given by more conventional high sulphur, accelerated systems when compounding for:

(a) retention of tensile properties on ageing at elevated temperatures

(b) low compression set under static or dynamic loads at ambient or elevated temperatures

(c) resistance to reversion of tensile properties during cure, and

(d) curing at high temperatures (up to 200°C). The reversion resistance at 200°C of compounds cured by EV systems may be five to ten times greater than that of conventionally cured compounds when measured in terms of curometer time to 10% reversion.

Table 9.7 illustrates the comparison of a particularly safe, fast curing conventional compound 13 with two of the fastest curing EV compounds 14, 15 so far investigated which also have adequate processing safety on the Daniels machine under the given conditions. Compared with these compounds, mix 16 has a somewhat slower cure rate at 200°C but offers greater processing safety (see curometer data).

Table 9.7 also illustrates some of the best results that have been achieved in terms of short cures with the barrel temperature restricted to 90°C and the mould to 200°C. For example, with the conventional compound 13, the thin side wall cures in 10 sec. and the 1 cm. thick base cures in 45-60 sec. Excellent reversion resistance may be obtained in up to 4 min. at 200°C using reversion resistant EV compounds which give side wall cures in 45 sec. and base cures in 120 sec. For example, there is virtually no loss in tensile strength or 300% modulus of the side wall while the base is curing.

9.5.5 EFFECT OF EXTENDERS ON INJECTION TIME, INJECTION TEMPERATURE AND CURE TIME

Barrel temperatures were raised to close to the highest temperature (125°C) that the conventional compound 17 (Table 9.7) could tolerate without scorching in the barrel. The average injection temperature for this compound was 158°C and the 1 cm. thick base cured to 95% of full

Table 9.6 Effects of varying cure system and fillers

Mix	1	6	7	8	9	10	11	12
RSS1 (yellow circle)	100	100	100	100	100	100	100	100
SRF black (Dixie 20)	50	50	50	50	100	—	—	—
MPC black						95		
FEF black							100	
HAF black								50
Dutrex R	5	5	5	5	10	5	5	5
MRX						3	5	
Pine tar								
Antioxidant 4010	1	1	1	1	1	2	2	1
Flectol H								
Zinc oxide	5	5	5	2	5	5	5	5
Stearic acid	2	2	2	2	2	3	2.5	2
Sulphur	2.5	2.5	2.5	—	2.5	3	2.5	2
CBS (Santocure)	0.5	—	0.5	—	0.5	0.8	1	1.2
Retarder								1
Santocure MOR	—	0.55	—					
TMTD	—	—	0.5	3	—	—	—	0.3
Mooney viscosity, ML3, 120°C	59	64	62	55	82	62	74	87
Mooney scorch, 120°C T5 min.	26	18	9	9	19	20	28	10

Injection moulding (Barrel 90°C)

Mould temperature °C	180	180	180	180	180	200	200	180
Nozzle diameter in.	5/32	5/32	5/32	5/32	5/32	5/32	5/32	1/8
Mean injection time, sec.	3·8	5·0	4·2	4·2	5·8	1·8	2·1	10·3
Injection temperature °C	118	120	119	117	115	96	100	128
Injection pressure (max.) psi	16 000							
Vulcanizate properties								
Optimum cure for side wall sec.	45	90	30	45	45	45	45	30
Optimum cure for base sec.	180	180	60	180	180	180	90	90
Side wall (1·6 mm. thick) at optimum cure:								
Tensile strength kg./cm²	226	195	197	199	148	173	163	278
Elongation at break %	520	490	445	555	340	330	260	420
300% Modulus kg./cm²	82	85	118	67	135	151	—	195
Side wall at optimum hardness of base:								
Tensile strength kg./cm²	168	176	182	191	135	142	132	230
Elongation at break %	480	510	455	505	340	345	280	360
300% Modulus kg./cm²	69	67	111	74	124	129	—	178
Base (1·0 cm. thick):								
Hardness IRHD	54	51	66	53	70	91	85	75
Press cure at 153°C min*	10	15	10	10	10	15	10	10
Tensile strength kg./cm²	225	224	192	228	155	196	186	202
Elongation at break %	550	495	390	550	335	330	295	355
300% Modulus kg./cm²	94	98	138	71	138	190	—	168
Hardness IRHD	58	56	63	52	73	88	86	72

* Maximum modulus cure or the first cure to reach within 10% of it.

Base mix: RSS1 (yellow circle), 100: SRF black (Dixie 20), 50; Dutrex R, 5; Zinc oxide, 5; Antioxidant 4010, 1.

Mix	13	14	15	16
Flectol H	–	2	2	2
Stearic acid	2	3	2	3
Sulphur	2	0·25	0·33	–
CBS (Santocure)	1·2	–	5·0	–
TMTD	0·3	1	0·5	1
Santocure MOR	–	2·1	–	2·1
Sulfasan R	–	–	–	0·5
Retarder	1	–	–	–
Mooney viscosity, ML3, 120°C	58	54	56	54
Mooney scorch, 120°C, T5 min.	9·5	13	23	15
Wallace-Shawbury curometer 200°C:				
Time to 10% crosslinking sec.	21	26	24	34
Time to 90% crosslinking sec.	36	62	54	86
Time to 10% reversion sec.	80	>800	335	450

Injection moulding

Barrel 90°C; mould 200°C; nozzle diameter $^5/_{32}$ in.; injection pressure (max.) 16 000 psi.

Mean injection time, sec.	3·9	4·1	4·2	3·8
Injection temperature °C	118	118	120	118

Vulcanizate properties

Optimum cure for side wall sec.	10	45	60	60
Optimum cure for base sec.	60	120	120	180
Side wall (1·6 mm. thick) at optimum cure:				
Tensile strength kg./cm^2	254	238	216	230
Elongation at break %	480	520	530	565
300% Modulus kg./cm^2	117	96	83	67
Side wall at optimum hardness of base:				
Tensile strength kg./cm^2	193	231	214	214
Elongation at break %	410	550	550	550
300% Modulus kg./cm^2	98	96	80	73
Base (1.0 cm. thick):				
Hardness IRHD	63	53	53	51
Press cure at 153°C min.*	5	25	10	25
Tensile strength kg./cm^2	211	215	222	215
Elongation at break %	450	520	470	550
300% Modulus kg./cm^2	114	87	99	69
Hardness IRHD	65	56	56	55

* Maximum modulus cure or the first cure to reach within 10% of it.

hardness in 43 sec. Table 9.8 shows that under these conditions dilution of NR by Shell polyisoprene (Cariflex IR305), reclaim and oil, and use of a process aid afford only a slight advantage in improved (shorter) injection time and that this is achieved at the cost of a lower injection temperature and a markedly longer cure time.

Table 9.9 Effect of white fillers

Base mix: RSS1 (yellow circle), 100; zinc oxide, 5; Nonox EXN, 1; CBS (Santocure), 0·5; sulphur, 2·50

Mix	24	25	26	27	28	29
Stearic acid	1	2	2	2	2	2
Calofort S	72	–	–	–	–	–
Devolite	–	72	120	–	–	–
Manosil AS 7	–	–	–	54	30	–
Mooney viscosity ML3, 120°C	59	42·5	37	110	80	46
Mooney scorch 120°C, T5, min.	22	28	40·5	11	20	42

Injection moulding data

Mould temperature °C	180					
Injection pressure (max.) psi	16 000					
Nozzle diameter in.	$^5/_{32}$					
Barrel temperature °C	125	125	126	114	117	111
Injection temperature °C	156	147	139	150	152	139
Extrudate temperature °C	139	137	129	127	125	129
Injection time sec.	2·0	1·5	1·4	3·5	2·3	1·5

Vulcanizate properties
1 cm. thick base
Hardness IRHD

Cure time	40 sec.	50D	45P D	–	56D	49P D	–
	50 ,,	51G	45P D	–	58D	50P D	–
	60 ,,	52	56P D	–	58G	49P D	31P D
	70 ,,	52	60G	–	59	52G	39G
	90 ,,	52	61	64P D	62	52	40
	120 ,,	52	61	67G	61	53	40

Dunlop resilience 21°C %						
(optimum cure)	84	83	74	74	81	89

Press cures–10 min. at 153°C:
0·8 cm. thick sheet

	24	25	26	27	28	29
Hardness IRHD	52	61	70	58	51	42
Dunlop resilience, 21°C	81	80	78	76	82	91

Side wall (1·6 mm. thick):
500% Modulus kg./cm.²

Cure time	30 sec.	124	–	–	121	112	–
	40 ,,	105	167	–	140	124	–
	50 ,,	88	–	–	144	92	–
	60 ,,	80	130	–	138	90	46
	90 ,,	77	118	143	144	87	44

	24	25	26	27	28	29
Tear strength at 150°C	30	40	90	40	40	60
Mean of maxima kg./mm.	0·47	0·19	0·18	1·00	0·47	0·08

Press cure–10 min. at 153°C:
0·2 cm. thick sheet

500% Modulus kg./cm.²	120	175	–	132	122	56

0·1 cm. thick sheet
Tear strength at 150°C

Mean of maxima kg./mm.	0·49	0·20	0·27	1·58	0·32	0·16

P refers to porosity and D to distortion due to undercure. G refers to good appearance.

Table 9.8 Effect of extenders on injection time, injection temperature and cure time

Base mix: Flectol H, 2; Antioxidant 4010, 1; zinc oxide, 5; cbs (Santocure), 0·5; sulphur, 2·5.

Mix	17	18	19	20	21	22	23
RSS1 (yellow circle)	100	80	50	–	80	80	100
Cariflex IR305	–	20	50	100	–	–	–
WT Reclaim	–	–	–	–	40	–	–
Sundex 8125 oil	–	–	–	–	–	21·6	–
Dutrex R	5	5	5	5	5	–	5
Actoplast	–	–	–	–	–	–	4
Stearic acid	2	2	2	2	2	2	1
SRF black (Dixie 20)	50	50	50	50	30	50	50
Mooney viscosity ML3, 120°C	55	57	64·5	17·5	37·5	30	49
Mooney scorch 120°C, T5, min.	28	31	30	49	32·5	36	34
Curometer 180°C							
Time to 90% crosslinking sec.	84	93	96	99	102	96	93
Injection moulding data							
Barrel temperature °C	125						
Mould temperature °C	180						
Nozzle diameter in.	5/32						
Injection pressure (max.) psi	16 000						
Injection temperature °C	158	149	147	141	146	140	150
Extrudate temperature °C	138	–	136	–	136	135	135
Injection time sec.	1·7	1·5	1·4	1·3	1·4	1·4	1·8

Vulcanizate properties—1 cm. thick base
Hardness IRHD:

Cure time	40 sec.	56G	51D	51D	–	45P D	44P D	51D
	50 ,,	57	52G	52D	–	47P D	46P D	54G
	60 ,,	58	53	52G	48P D	47P D	48P	56
	70 ,,	–	56	54	51G	49P	48P	57
	80 ,,	–	57	54	51	50P	51G	57
	90 ,,	59	57	56	50	50G	51	59

102

Time to 95% of full hardness sec.*	43	57	71	66	82	85	61
Dunlop resilience, 21°C, % (optimum cure)	77	80	81	76	72	75	79
Press cures—10 min. at 153°C, 0.8 cm. thick sheet:							
Hardness IRHD	59	58.5	56.5	54	51	54	57
Dunlop resilience 21°C %	81	79	81	81	71	77	77
Side wall (1.6 mm. thick)							
300% Modulus kg./cm²							
Cure time 30 sec.	93	75	51	—	57	—	86
40 "	95	79	66	—	61	77	90
50 "	91	76	63	—	58	77	84
60 "	89	69	68	42	58	69	80
90 "	79	71	66	47	59	72	79
Cure time at 180°C sec.	40	40	40	70	40	40	40
Tear strength at 150°C							
Mean of maxima kg./cm².	0.44	0.37	0.30	0.37	0.52	0.52	0.60
After ageing for 3 days at 100°C							
Tensile strength kg./cm²	145	130	116	59	65	86	139
Elongation at break %	380	385	380	220	265	290	445
100% Modulus kg./cm²	24	21	20	20	13	20	19.5
300% Modulus kg./cm²	112	95	87	—	—	—	85
Press cures—10 min. at 153°C							
0.2 cm. thick sheet							
300% Modulus kg./cm²	99	84	74	69	62	80	84
0.1 cm. thick sheet							
Tear strength at 150°C							
Mean of maxima kg./mm.	0.38	0.44	0.37	0.21	0.63	0.47	0.65
After ageing for 3 days at 100°C							
0.2 cm. thick sheet							
Tensile strength kg./cm²	144	146	153	96	73	95	151
Elongation at break %	325	425	495	350	275	310	470
300% Modulus kg./cm²	132	105	93	82	—	90	89

P refers to porosity and D to distortion due to under cure.   G refers to good appearance.   *   Graphically determined.

For example, 20% dilution of NR by Cariflex IR305 (mix 18) increases the cure time of the base by 32% to 57 sec. and similar dilution by reclaim and oil delays cure to 80-90 sec. respectively. The side walls which are thinner and capable of receiving quicker heat transfer from the mould are not so markedly affected by dilution. Cures shorter than 30 sec. could not be examined because the base was not strong enough to permit proper extraction of the whole moulding.

9.5.6 EFFECT OF WHITE FILLERS

Table 9.9 shows that mixes containing a volume loading, equivalent to that of 50 phr carbon black, of precipitated chalk (Calofort S), Devolite clay and precipitated aluminium silicate (Manosil AS 7) may be satisfactorily moulded. Calofort S and Devolite clay mixes afforded high injection temperatures similar to those given by SRF black. The Manosil AS 7 mix was more scorchy and gave such a high heat build-up in the nozzle that the barrel temperature had to be reduced, even when a lower volume loading was employed. A mix with a higher volume loading (120 phr) of Devolite gave lower injection temperatures and cured slowly in comparison with the mix having a lighter loading. Table 9.9 also includes a conventional gum mix.

Mixes which have rather low tear strength, such as the highly loaded clay mix, are inclined to leave a very fine flash around the parting mould surfaces and this tends to become sticky if it is not wiped off quickly.

9.5.7 USE OF THE PECO 21TS INJECTION MOULDING MACHINES TO MAKE THIN SHEET MOULDINGS

The Peco 21TS machine was heated by hot water and controlled close to 90°C and thus the working conditions were somewhat similar to those with Daniels machine when its barrel was controlled at 90°C. The screw speed and the injection pressures on the Peco machine were higher than those used with the Daniels machine and the mould was simpler in form and smaller.

The Peco 21TS machine was used to make a comparison of the curing behaviour of two conventional natural rubber compounds and two EV compounds also examined on the Daniels machine (see Table 9.10 for detail). Thin sheet mouldings (10 cm. diameter, 0·28 cm thick weighing 25 g. Fig. 9.13 Plate 10) of both conventional and EV compounds cured in 15-20 sec. at 200°C. Figures 9.14-9.17 confirm data obtained with the Daniels machine that, at their optima, the tensile properties of the injection moulded vulcanizates are similar to those obtained by press curing at 153°C, and they also illustrate the excellent reversion resistance of the EV compounds.

9.5.8 PHYSICAL PROPERTIES OF VULCANIZATES

Comparison of data for injection moulded vulcanizates and press cured vulcanizates made from identical mixes may be made in Tables 9.6-9.9 and Figs 9.14-9.17. Under the variety of conditions employed it may be concluded that, although 300% modulus of injection moulded vulcanizates cured by conventional systems may be 5-10% lower, than that of press cured vulcanizates, the physical properties of injection

Table 9.10 Conditions for moulding with the Peco 21TS machine

Base mix: RSS1, 100; SRF black, 50: Dutrex R, 5; Flectol H, 2; Antioxidant 4010, 1; zinc oxide, 5; stearic acid, 2.

	Conventional		EV	
Mix no.	17	13	14	15
Sulphur	2·5	2·0	0·25	0·33
CBS (Santocure)	0·5	1·2	–	5
TMTD	–	0·3	1·0	0·5
Santocure MOR	–	–	2·1	–
Retarder	–	1	–	–
Mooney viscosity ML3, 120°C	32	48·5	46·5	42·5
Mooney scorch, 120°C, T5, min.	27·5	14·3	19·5	16·5

Injection moulding machine data
Nozzle diameter, 0·1 in.
Injection: line pressure, 1 000 psi
  (material pressure 15 000 psi)
Injection boost: line pressure, 1 200 psi
  (material pressure 18 000 psi)
Screw drive pressure, 1 000 psi
Screw speed 120 rpm
Moulding 0·28 cm. thick sheet
Screw back pressure, 120 psi;
  injection boost time, 0·5 sec.;
  injection pressure time, 2 sec.

Barrel temperature °C	93	90	93	92
Mean platen temperature °C	204	198	210	203
Mean injection time sec.	2·3	2·5	2·6	2·2
Injection temperature °C	125	130	137	128

Vulcanizate properties—see Figs 9.14–9.17
  and 9.35

mouldings are, at their optima, similar in tensile strength, elongation at break, tear strength, hardness, resilience and oven ageing resistance.

If tensile strength is consistently lower than expected it can sometimes be traced to surface defects due to scorch during flow through the mould.

Dorko, Timar and Walker [38] found that fast injection times lead to physical properties equal to, or better than, those obtained by compression moulding. Bockmann [39] obtained better dynamic (heat build-up) and static properties from injection and transfer moulded parts than from compression moulding.

9.6 PREPARATION OF THICKER TEST PIECE MOULDINGS AND MOULDINGS OF COMMERCIAL INTEREST

9.6.1 INTRODUCTION

NR is widely used in engine mountings and shock absorbers and, as articles of this nature involve relatively thick sections particular attention has been paid to test piece mouldings which simulate a shock absorber.

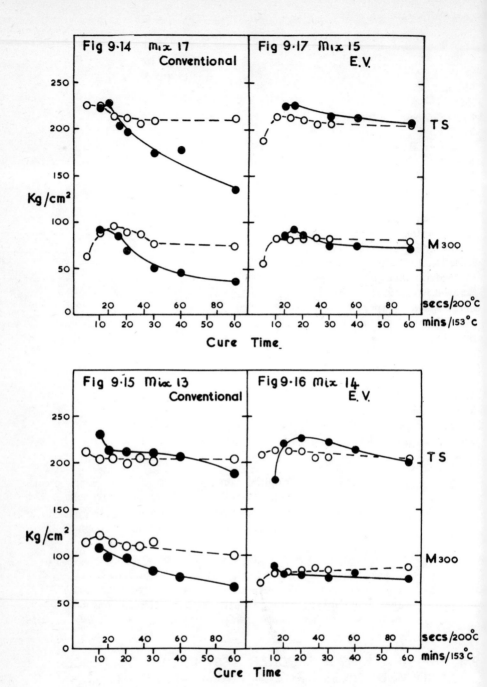

Figs 9.14-9.17. Comparison of the tensile strength and the 300%
modulus of 0·28 cm. thick vulcanizates of conventional mixes 17, 13
and EV mixes 14, 15: (●—●—●) when injection moulded and cured at
200°C; (O—O—O) when press cured at 153°C. (Peco 21TS machine.
See also Table 9.10.)

9.6.2 EV SYSTEMS

The Peco 21TS [24] machine was used under conditions detailed in Table 9.11 for the preparation of mouldings (Fig. 9.1) from two conventional NR compounds and two NR compounds cured with reversion resistant efficient vulcanization (EV) curing systems [35, 36, 37] The mouldings were injected at about 140°C and they

Table 9.11 Base mix: RSS

Base mix: RSS1, 100; SRF black, 50; Dutrex R, 5; Flectol H, 2: Antioxidant 4010, 1; zinc oxide, 5; stearic acid, 2.

	Conventional		EV	
Mix no.	17	13	14	15
Sulphur	2·5	2·0	0·25	0·33
CBS (Santocure)	0·5	1·2	–	5
TMTD	–	0·3	1	0·5
Santocure MOR	–	–	2·1	–
Retarder	–	1	–	–
Mooney viscosity ML3, 120°C	32	48·5	46·5	42·5
Mooney scorch, 120°C, T5 min.	27·5	14·3	19·5	16·5

Injection moulding machine data
Nozzle diameter, 0·1 in.
Injection: line pressure, 1 000 psi
    (material pressure, 15 000 psi)
Injection boost: line pressure, 1 200 psi
    (material pressure, 18 000 psi)
Screw drive pressure, 1 000 psi
Screw speed, 120 rpm
Moulding 2·1 cm thick cylinder
Screw back pressure, 100 psi;
    injection boost time, 4 sec.;
    injection pressure time, 10 sec.

Barrel temperature °C	90	88	89	88
Mean platen temperature °C	201	200	200	200
Mean injection time sec.	0·64	1·9	1·2	1·4
Injection temperature °C	136	142	141	140

Vulcanizate properties—see Figs 9.18–9.21

cured in 60–90 sec. at 200°C. Figures 9.18–9.21 show changes in mean surface hardness and mean internal hardness with cure time and also indicate the presence of porosity (P) and or distortion (D) due to undercure. It was necessary for these mouldings to remain in the mould for sufficient time to prevent porosity even though, after only 30 sec. in the mould, they had the heat capacity to come to a full state of cure after being ejected. The figures show that the conventional compounds have the advantage of curing somewhat faster than the particular EV compounds examined here. However, the EV compounds have the advantage that they exhibit less surface reversion and therefore have a more uniform hardness throughout their cross-section.

It may be concluded that, although they are 'peaky', conventional curing systems may be used satisfactorily, in short cures at 200°C. However, with present injection moulding machines, injection into a mould takes place well below mould temperature and curing depends to some extent on relatively slow heat transfer from the mould. Under these circumstances, especially at temperatures as high as 200°C, and where 'peakiness' and reversion may be expected, EV systems [35, 36, 37] permit the curing of thicker articles than would be possible with conventional curing systems.

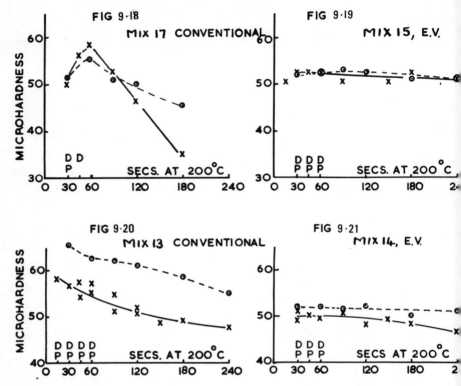

Figs. 9.18–9.21. Comparison of the internal (O–O–O) and surface (X–X–X) hardness of 2·1 cm. thick mouldings of conventional compounds (17, 13) and EV compounds (14, 15) cured at 200°C. P indicates porosity and D indicates distortion due to undercure. See also Table 9.11.

9.6.3 SOLID MOULDINGS 1·5 in. (3·8 cm.) THICK

Bearing in mind the conclusions drawn earlier and in the previous section it was felt desirable to obtain the highest possible injection temperature by increasing barrel temperatures to their maximum consistent with freedom from scorch. Mould temperatures in this section of the work do not exceed 180°C because, above this temperature, some mouldings scorched at the surface while the mould was filling.

The Daniels Edgwick 45-SR machine [25] was used in conjunction with a frustum cone mould (Fig. 9.22 Plate 11), designed by RAPRA [34] to make use of the maximum shot volume of the machine. Vacuum leads were drilled into a thin flash groove at the base of the mould and the mould was evacuated by a laboratory vacuum pump for 15 sec.

after each mould closure. Necessity for evacuation was demonstrated when, occasionally, the flash sealed off the pump and caused very serious air trapping.

Table 9.12 shows that 60 sec. cures are possible with a conventional mix 17 and two EV mixes when the barrel temperature is 120-125°C, injection temperatures are 140-150°C and the mould is 180°C. The scorchier and potentially faster curing conventional mix 13 scorched when the barrel was held at 110°C and stock emerging from the head without the nozzle reached 136°C. This illustrates the advantage of using a fairly safe cure system because inability to raise the barrel temperature above 105°C resulted in cure time of 90-120 sec. for this mix. Table 9.12 also shows that mould temperatures less than 180°C lead to unnecessarily long cure times. Further mix variations were examined as can be seen in Tables 9.13 and 9.14.

Surface and internal hardness measurements (the latter made on cut mouldings) show (Tables 9.12-9.14) that, although the shortest (45 sec.) cures capable of being extracted from the press are sufficiently undercured to develop porosity after extraction, they do have the heat capacity to come to full hardness while cooling down outside the mould. Needle probe measurements made on mouldings immediately after removal from the mould show that 60 sec. cures having maximum internal mould temperatures 147-152°C may remain for at least 10 min. above 140°C at their centres and it is quite understandable that full cure should be reached during cooling.

Internal and surface hardness measurements on any given moulding are virtually the same for conventional compounds (mixes 17, 30, 31, Table 9.13, Fig. 9.23) injected at 145-150°C into a mould at 180°C and there is no evidence of reversion in cures up to 2 min. This is in contrast to the reversion observed with conventional compounds at 200°C as detailed in Table 9.11 and Figs 9.18 and 9.19 and, bearing in mind the short cures obtained, illustrates the advantage of using higher barrel temperatures and the lower mould temperatures.

Three conventional mixes (17, 30, 31, Table 9.13, Fig. 9.23) containing different antioxidant systems were overcured (4 min.) and maximum internal mould temperatures reached 160-165°C but, in spite of this, no serious internal or external reversion, as in Figs 9.18, 9.19, was observed with the mould at 180°C. Neither was there any significant difference between the vulcanizates in their states of cure after 4 min. when the mixes were protected by either 1 phr PBN, 1 phr Flectol H or 2 phr Flectol H with 1 phr Antioxidant 4010.

Table 9.13 also shows that, when NR in the compound is diluted by Shell polyisoprene or oil, or when it is completely replaced by Goodyear's Natsyn 200, only trivial advantages in reduced mould filling time are obtained. These advantages are, however, only obtained at the cost of lower heat build-up during passage through the screw and through the nozzle and at the cost of longer cure time. The effects of this dilution on cure times are more marked when dealing with thicker sections as can be seen by comparison with data reported for the 1 cm. thick beaker base in Table 9.8 of this chapter.

Table 9.14 shows that mixes containing other fillers such as HAF and

Table 9.12 Variation of barrel and mould temperatures

	Conventional					EV	
Mix	17	17	17	13	14	15	15
Mooney viscosity ML3, 120°C	58.5	58.5	58.5	56	57	51	51
Mooney scorch 120°C, T5 min.	23	23	23	11	15	12	12
Mould temperature °C	180	160	180	170	180	170	180
Injection pressure (max.) psi	16 000						
Nozzle diameter in.	5/32						
Barrel temperature °C	100	125	122	105	124	120	120
Injection temperature °C	142	—	149	144	150	—	141
Injection time sec.	2.9	2.5	—	—	2.1	1.5	1.6
Vulcanizate properties							
Micro-hardness IRHD							
Surface							
Cure time 45 sec.	56.5P D	—	60P D	66P D	—	—	56.5G
60 "	58P	57P	60G	67P D	55.5D	53P D	58
90 "	60G	60.5G	60.5	65.5G	55.5G	56.5G	56
120 "	59	61	60	65.5	55	56	—
150 "	—	—	58.5	66.5	—	—	—
Internal							
Cure time 45 sec.	—	—	58.5	—	57.5	55.5	59
60 "	59.5	58	58	62	57	56	58
90 "	58	59.5	60	63	57	58.5	—
120 "	—	—	56	62	—	—	—
150 "	—	—	59	—	—	—	—
Press cures at 153°C							
Hardness IRHD							
Cure time 10 min.	59.5	59.5	59.5	64	56.5	56.5	56.5

In these tables P indicates porosity, and D distortion due to undercure. G indicates good appearance.

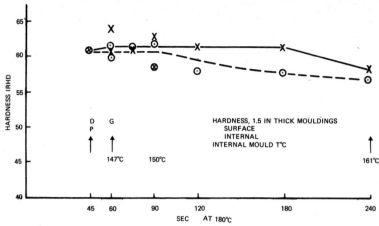

Fig. 9.23. Comparison of the internal (O–O–O) and surface (X–X–X) hardness of 1·5 in. thick mouldings of conventional mix 31 cured at 180°C. P indicates porosity, D distortion and G, good appearance.

MT black at a 50 phr loading and white fillers at equivalent volume loading can also be satisfactorily injection moulded in thick sections. Mixes including a gum stock (mix 29) which gave high injection temperatures (146°C) cured satisfactorily in 60–90 sec. A clay mix and a silica mix which injected at lower temperatures required longer curing times The Manosii AS 7 mix was too scorchy in the barrel to be handled at barrel temperatures above 115°C, but it injected at 147°C and cured in 60 sec.

9.6.4 USE OF THE FARREL-BRIDGE MACHINE TO MAKE THICK MOULDINGS

The Farrel-Bridge Model R60-350 rubber injection moulding machine [11] was used for preparation of solid wheelbarrow wheel mouldings weighing 1150 g. (SG 1·1) and having a 1·9 in. diameter cross-section (Fig. 9.24 Plate 11).

The machine differs from the Daniels and Peco machines in being operated on a simple hydraulic ram principle. The ram drives the rubber compound through a heating chamber containing a torpedo, through a nozzle which gives a heat build up of 20°C and then through heated runners which form part of the mould and which give a further heat build up. Table 9.15 provides further details and shows that full cures of an EV compound can be obtained in a 2 min. nominal cure time when the centre of the moulding reaches 151°C.

In conventional moulding the necessary cure time for the wheelbarrow wheel is 20 min. and therefore the injection moulding process may be 8–10 times faster.

9.6.5 USE OF THE FOUR-STATION DESMA 905 MACHINE TO MAKE SOLID MOULDINGS 4 in. (10 cm.) THICK

In production of exceptionally large, thick-walled products such as a rubber buffer, it may be necessary to fill a mould having greater

Table 9.13 *Variation of protective system and extenders*

Base mix: SRF black, 50; zinc oxide, 5; stearic acid, 2; sulphur, 2·5; CBS (Santocure), 0·5.

Mix	30	31	17	18	19	32	22
RSS1 (yellow circle)	100	100	100	80	50	—	80
Cariflex IR305	—	—	—	20	50	—	—
Natsyn 200	—	—	—	—	—	100	—
Sundex 8125	5	5	5	5	5	5	21·6
Dutrex R	5	5	5	5	5	5	—
PBN	1	1	—	—	—	—	—
Flectol H	—	—	2	2	2	2	2
Antioxidant 4010	—	—	1	1	1	1	1
Mooney viscosity ML3, 120°C	53·5	57	52·5	59·5	51·5	56·5	37
Mooney scorch 120°C, T5 min.	31	36	32	32·5	36	37	34
Mould temperature °C	180						
Injection pressure (max.) psi	16 000						
Nozzle diameter in.	5/32						
Barrel temperature °C	120	120	122	121	121	120	123
Injection temperature °C	145	145	149	143	138	134	140
Extrudate temperature °C	137	137	140	133	132	124	126
Injection time sec.	1·6	1·5	1·7	1·5	1·6	1·6	1·7

Vulcanizate properties

Micro-hardness IRHD

Surface

Cure time							
50 sec.	59·5P D	61P D	59G	P D	—	P D	54P D
60 ,,	60·5D	61G	59	56·5P D	43·5P D	P D	54G
90 ,,	60G	61	60	57·5P D	43·5P D	51P D	55
120 ,,	61·5	61·5	60·5	58P D	52·5P D	51P D	55·5
150 ,,	—	—	—	57·5G	54·5P		56·5
180 ,,	58	61·5	58·5	—	54·5P	56·5G	—
240 ,,	57·5	58·5	57	—	54·5G	57	

Internal

Cure time							
50 sec.	61	61	57·5	57·5	—	—	—
60 ,,	59	61·5	55·5	58·5	50·5	47	55
90 ,,	60·5	60·5	56	59	48	34	55
120 ,,	59·5	58	57·5	58·5	53·5	50	58
150 ,,	—	—	—	60	53·5	52	55·5
180 ,,	58·5	58	55	—	55·5	52·5	56
240 ,,	57·5	57	53·5	—	56·5	54	—

Press cures at 153°C

Hardness IRHD

Cure time 10 min.	58·5	57·5	57·5	58	56·5	57·5	54·5

P refers to porosity and D to distortion due to under cure.   G refers to good appearance.

113

Table 9.14 Variation of fillers at equal volume loading and a gum mix

Mix	33	34	24	25	27	35	29
RSS1 (yellow circle)	100	100	100	100	100	100	100
Zinc oxide	5	5	5	5	5	5	5
Stearic acid	2	2	1	2	2	2	2
CBS (Santocure)	0·5	0·5	0·5	0·5	0·5	2	0·5
Sulphur	2·5	2·5	2·5	2·5	2·5	3	2·5
Antioxidant 4010	1	1	—	—	—	—	—
Flectol H	2	2	1	1	1	1	1
Nonox EXN	—	—	—	—	—	—	—
Dutrex R	5	5	—	—	—	—	—
HAF black	50	—	—	—	—	—	—
MT black	—	50	—	—	—	—	—
Calofort S	—	—	72	72	—	—	—
Devolite Clay	—	—	—	—	54	—	—
Manosil AS 7	—	—	—	—	—	—	—
Manosil VN 3	—	—	—	—	—	54	—
Digol	—	—	—	—	—	2	—

Mooney viscosity ML3, 120°C	65	47	63	42	110	105	46
Mooney scorch 120°C, T5 min.	18	39	31	44	11	26	42
Mould temperature °C	180						
Injection pressure (max.) psi	16 000						
Nozzle diameter in.	5/32						
Barrel temperature °C	121	122	122	122	115	122	120
Injection temperature °C	146	146	146	137	147	137	146
Injection time sec.	1·5	1·6	2·3	1·4	2·6	2·5	1·4
Vulcanizate properties							
Micro-hardness IRHD							
Surface							
Cure time 60 sec.	66G	50P D	53P D	–	60·5G	–	39D
90 „	68	51G	54G	P D	61·5	P D	41G
120 „	67	53	54	P D	63	–	40
150 „	–	52	54	59P D	–	77D	–
180 „	–	–	–	58·5P	63	–	–
240 „	–	–	–	–	–	82G	–
Internal							
Cure time 60 sec.	69	52	53·5	–	56	–	38
90 „	69	51·5	56·5	–	57	–	37
120 „	67·5	52·5	55·5	57·5	60	67	37
150 „	–	52	51·5	60	–	73	–
180 „	–	–	–	–	61	–	–
240 „	–	–	–	–	–	–	–
Press cures at 153°C							
Hardness IRHD							
Cure time 10 min.	65	55	52·5	61	58	83	42

Mix no.	15		
Mooney viscosity ML3, 120°C	52		
Mooney scorch 120°C, T5 min.	21		
Injection moulding data			
Nozzle diameter in.	0·269		
Clamping pressure psi	3 000		
Injection pressure psi	2 000 (Material pressure 25 000 psi)		
Mould temperature °C	190		
Heating chamber temperature °C	113		
Nozzle temperature °C	107		
Injection temperature °C	127		
Injection time (set) sec.	12		
Sprue break time sec.	25	25	40
Cure time sec.	60	120	180
Total cure time* sec.	73	133	208
Injection time observed sec.	15	12	12
Maximum internal mould T °C	142	151	161
Hardness IRHD	58	59	59
Appearance	porous	good	good

* Total cure time = cure time + sprue Break time − injection time.

capacity than the shot volume of the injection moulding machine. In this case it is possible, either to inject several times at the necessary plasticizing time interval, or to screw in the whole volume of rubber at a relatively low pressure through wide runners. The latter technique has been called 'intrusion' or 'flow moulding'. Jäkel [40] has shown that a 5 200-cc. (7·9 in. diameter) solid rubber buffer can be intrusion moulded with a 240-cc. shot capacity Eckert and Ziegler machine and cured in 30 min. compared with 10 hr. for conventional compression moulding.

The four-station Desma 905 machine was used to make test piece mouldings from a bowling pin mould (Fig. 9.25 Plate 11) of volume 2·14 litres and maximum cylinder diameter 9·8 cm. (3·85 in.). The shot volume of this machine is 1·7 1. (80% of the moulding volume) Trials were made with procedures using double injection, complete intrusion and a combination of injection of 1·7 1. followed by intrusion of the remaining volume. Double injections caused some scorch of the first injected volume on the mould surface (at 180°C). Complete intrusion was more time consuming than injection followed by intrusion and so the latter process was adopted as the most favourable. Table 9.16 shows that when screw speeds of 100 rpm (back pressure 15 kg./cm²) were used with NR/50 HAF black compounds plasticizing times were reduced to 21–26 sec. and injection times to 15 sec. (7 sec. to inject and 8 sec. to screw in the remainder). Injection temperatures (samples prior to making the moulding) reached 158–165°C. Under these conditions the 9·8 cm. thick bowling pin samples cured all the way through in 6 min. with the mould at 180°C. Cures lasting 4 and 5 min. were only very slightly porous (Fig. 9.26 Plate 12) in the centre leaving the hope that the cure time could still be

Table 9.16 Four-station Desma 905 Machine moulding conditions for curing bowling pin, ski-boot last and pram tyre mouldings

Compound (see below)	36	37	37	38	39	40	39	39	40
Mould	Bowling pin	Bowling pin	Bowling pin	Bowling pin	Bowling pin	Bowling pin	Ski-boot last	Pram tyres	Pram tyres
Cylinder T, zone 1 °C	115	105	108	127	90	100	98	100	90
Cylinder T, zone 2 °C	115	110	112	137	95	100	105	110	100
Cylinder T, zone 3 °C	120	110	110	130	95	100	105	110	100
Nozzle T, °C	122	115	115	105	85	90	93	97	90
Screw coolant T, °C	90	90	90	90	90	90	90	90	90
Injection T, °C	158	162	165	145	145	142	142	142	139
Mould T, °C	180	180	180	180	180	180	185	180	183
Screw speed rpm	100	100	100	126	63	100	63	63	63
Screw motor amp	80	90	88	78	60	70	60	60	46
Back pressure kg./cm²	15	15	15	15	15	12	15	15	12
Injection pressure kg./cm²	1 000	1 000	1 000	1 000	900	1 000	900	900	900
Plasticizing time sec.	26	21	22	25	–	19	–	27	30
Injection time sec.	15	19	18	15	35	15	10	12	11
Vulcanization time min.	4	5	6	10	8	9	8	2·25	4·5
Internal mould T (1 in. inside surface °C)	148	153	153	161	156	145	142	–	–
Appearance of moulding	very slightly porous	very slightly porous	slightly porous	good	good	good	good	good	good

Compounds in Table 9.16

Compounds	36	37	38	39	40
SMR5 Heveacrumb	100	–	53·6	100	100
RSS1	–	100	13·4	–	–
Intene 55NF	–	–	–	5	–
Dutrex R aromatic oil	4	4	31	–	–
Petrofina 2069 paraffinic oil	–	–	–	–	–
HAF black	50	–	–	–	–
ISAF black	–	50	55	–	–

	39 Bowling pin	39 Ski-boot last	39 Pram tyres	40 Pram tyres
SRF black	–	–	–	50
Flectol H	2	2	2	1
AO 4010	–	–	–	1
Nonox ZA	2	2	2	5
Zinc oxide	5	5	5	5
Stearic acid	2	2	2	2
Vulcatard A	1	1	1	–
Sulphur	2·5	2·5	2·0	2·5
CBS (Santocure)	0·5	0·5	0·8	0·5

117

significantly reduced by obtaining only a slightly higher injection temperature.

Temperatures measured by a needle probe placed 1 in. inside the surface of the moulding reached 148°C for a 4 min. cure (1 min. after extraction) but were not higher than 153°C for 6 min. cures indicating that pre-injection samples injected into atmospheric pressure attained a higher temperature than the mass injected and intrusion moulded against the pressures of mould filling.

A sample which reached 153°C 1 min. after ejection was still 142°C 1 in. under the surface after a further 10 min. cooling. It is therefore quite understandable that such a thick cylinder should have sufficient heat capacity to cure after 6 min. in the mould and yet not suffer any undue reversion.

Hardness across the diameter of the sample was shown to be constant except for the outside skin about 0·5 cm. thick where hardness dropped only a few degrees.

An oil extended NR compound (67 NR/31 Petrofina 2069 oil/55 ISAF black see Table 9.16) having a lower Mooney viscosity (ML3, 120°C, 40) and greater scorch resistance required a screw speed of 126 rpm for injection temperatures to reach 142-145°C. After 10 min. at 180°C the bowling pin was cured right through. Nine-minute cures were only slightly porous and 8 min. cures were badly blown.

Sample temperatures 1 in. below the surface, 1 min. after extraction were 149, 156 and 159°C for 8, 9 and 10 min. cures respectively.

The SRF black compound (NR/50 SRF black see Table 9.16) gave injection temperatures 142-145°C with a screw speed 63 rpm and required 8 min. for complete cure of the bowling pin.

A gum compound (see Table 9.16) cured in 9 min. when injection temperature reached 142-146°C with screw speed 100 rpm.

This data again emphasizes the importance of gaining high injection temperatures in the curing of thick objects. Injection temperatures near 150°C are almost twice as effective as temperatures near 140°C.

Another large volume moulding, a ski-boot last (2·06 l.) shown in Fig. 9.27 Plate 12, required injection followed by intrusion of the remaining volume. The SRF black compound injected in 10 sec. at 142°C and cured in 8 min. at a mould temperature 185°C (Table 9.16). A set of four pram tyres moulded as shown in Fig. 9.28 Plate 12, having an overall shot volume 1·415 l. injected at 142°C in one shot in 12 sec. and cured in 135-150 sec. at 180°C. (Table 9.16.)

The Desma 905 machine was also used in the manufacture of a pair of sealing rings (Fig. 9.29 Plate 13) moulded together as indicated in the diagram. Perfect mouldings were obtained with no sign of separation where the two streams of rubber combined to complete the ring opposite the injection point. Cures were obtained in 60-70 sec. with the two different compounds (Table 9.17) at the relatively modest mould temperature of 174°C. Other moulding details are given in Table 9.17.

The quartet of shoe sole and heel units moulded as in Fig. 9.30 Plate 13, cured in 2-2½ min. at 174°C as shown in Table 9.17. Injection temperatures were 135°C for the oil extended aluminium

PLATE 13

Fig. 9.29. Pair of sealing rings as moulded: larger ring, outside diameter 22 cm. (8·6in.); smaller ring, outside diameter 18 cm. (7·05in.); shot weights including sprue; mix 41, 142 g.; mix 42, 134 g. (Courtesy of NRPRA).

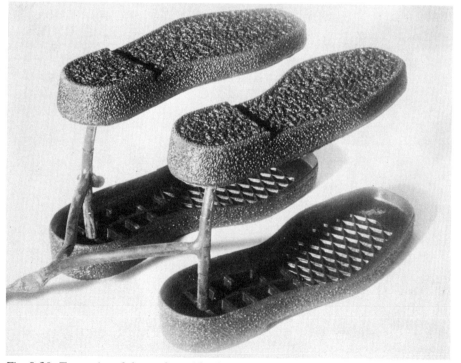

Fig. 9.30. Two pairs of shoe sole and heel units: larger, length, 23·8 cm. (9·4in.); smaller, length, 21·1 cm. (8·3in.); shot weights including sprue: mix 41, 540 g.; mix 42, 500 g. (Courtesy of NRPRA)

PLATE 14

PLATE 14

Fig. 9.31. (Right) The Stübbe plaster bowl mould; maximum diameter: 12·8 cm. (4 in.); side-wall, 3 mm. thick; base 4·5 mm. thick; thickest section across rim at base, 9·0 mm.; weight 117 g. of compound, SG 1·1. Sprue bush diameter, 4·5 mm. (courtesy: Stübbe).
Courtesy of NRPRA.

Fig. 9.32. (Left) A selection of blood test phial stoppers produced from two different natural rubber compounds. Both compounds illustrate the excellent flow properties and high production potential of NR in a complicated multi-cavity mould.
Courtesy of NRPRA.

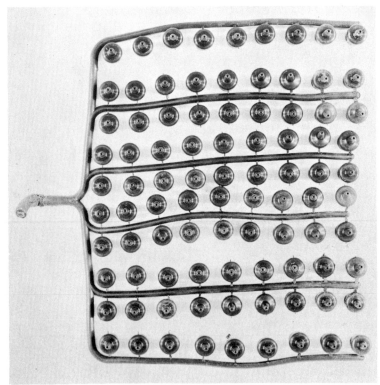

Fig. 9.33. A selection of mouldings illustrating the runner and gate system. (Courtesy of NRPRA).

Table 9.17 Injection moulding data, four-station Desma machine 905

	Sealing rings (Fig. 29 Plate)		Shoe sole and heel units (Fig. 30 Plate)	
Mix (see below)	41	42	41	42
Screw diameter mm.	90	–	–	–
Screw length L/D	12	–	–	–
Clamping force (max.) Mp	500	–	–	–
Screw speed rpm	61	61	50	61
Screw motor amp.	50	55	50	55
Screw back pressure kg./cm²	0	100	0	100
Cylinder temperature Zone 1°C	65	–	–	–
Zone 2°C	70	–	–	–
Zone 3°C	75	–	–	–
Interior screw temperature °C	70	–	–	–
Injection temperature °C	135	142	135	142
Mould temperature °C	174	182	174	174
Injection pressure kg./cm²	1 100	1 100	750	1 100
Injection time sec.	6	5	–	8
Cure time sec.	65–70	60	150	120

Mix	41	42
SMR 20	90	–
RSS1 (yellow circle)	–	100
Petrofina 2069 oil	10	5
Manosil AS 7	65	–
SRF black (Dixie 20)	–	50
Nonox WSP	1	–
Flectol H	–	1
Antioxidant 4010	–	1
Zinc oxide	5	5
Stearic acid	2	2
Sulphur	2·5	2·5
CBS (Santocure)	0·5	0·5
Mooney viscosity ML4, 100°C	75	55
Mooney viscosity ML3, 120°C	55	43
Mooney scorch 120°C, T5 min.	25	41

silicate compound and 142°C for the black compound.

The Stübbe S150/235 injection moulding machine was used in conjunction with a plaster bowl mould (Fig. 9.31 Plate 14). Table 9.18 illustrates the uniformity of the properties of tensile strips cut vertically from the bowl as the cycle time was reduced from 64 to 45·6 sec. by increasing screw speed and injection pressure.

The cycle time was defined by the operations: mould close, inject, cure, mould open—leaving out stripping the mould. The barrel heating fluid was controlled at 60–65°C because scorch in the barrel had been observed at 80°C when the screw speed reached 80 rpm. The mould temperature was held at 175°C because scorch had been observed during

Table 9.18 Injection moulding data single station Stübbe SI50/235 machine

Mix no. 42 (Table 9.17
Plaster bowl mould, Fig. 9.3)

Moulding serial no.	38	46	65	67	70	72	73	75
Barrel temperature °C	60	—	—	—	55	—	—	55
Mould temperature °C	175	—	175	—	—	—	—	172
Injection pressure atm.(line)	120	135	145	—	140	140	—	140
Injection pressure psi (material)	15 350	17 300	18 550	—	17 920	17 920	—	17 920
Hold pressure atm.	50	55	55	—	55	55	—	—
Screw speed rpm	50	80	75	75	75	75	75	75
Screw back time sec.	45	35	27	27	30	30	30	30
Injection time sec.	5	2·0	1·6	1·6	—	—	2·1	2·1
Vulcanization time sec.	54	43	42	42	31	31	30	30
Cycle time* sec.	64	52	58	58	48·7	48·7	45·6	45·6
Injection temp. °C		146						140
Physical properties of vulcanizates								
Tensile strength kg./cm²	192	198	194	200	199	189	199	189
Elongation at break %	535	540	535	520	520	515	520	495
M100 kg./cm²	14·5	15	13·5	16·5	16	15	16	17
M300 kg./cm²	79	80	80	90	86	84	88	92
M500 kg./cm²	177	176	183	187	186	182	188	
Hardness IRHD	56	57	54	56	55	55	56	58

* Cycle: close mould, inject, cure and open mould (stripping not included).

mould-filling at 185–190°C.

9.6.6 USE OF A NEW PISTON MACHINE WITH A MULTI-CAVITY MOULD

The Seidl SPA 1 BX-A machine, which is a development of former piston machines and offers improved preheating and plasticization in two stages by means of a pair of novel ball bearing packed plasticization cylinders, was successfully used in the manufacture of blood test phial stoppers (Figs 9.32 Plate 14 and 9.33 Plate 14).

Two NR compounds were injection moulded under conditions given in Table 9.19 using a mould designed to produce 90 blood test phial stoppers. The complete filling of the mould (Fig. 9.34 Plate 15) which consists of a complicated runner and gate system fed from one injection point at the parting surfaces of the mould, demonstrates the good flow properties and the high production potential of NR in injection moulding with multi-cavity moulds.

Table 9.19 Injection moulding data Seidl SPA 1 BX-A machine

Mould (Fig. 9.34 Plate 15); mouldings (Figs 9.32 Plate 14 and 9.33 Plate 14); mixes	43	44
Barrel temperature Zone 1°C	100	90
Zone 2°C	80	80
Zone 3°C	80	80
Zone 4°C	80	80
Injection temperature °C	130	136
Mould temperature °C	170	180
Injection time sec.	8	8
Injection pressure psi	18 000	18 000
Mould locking pressure tons	330	330
Nozzle diameter mm.	5	5
Cure time sec.	120	35
Micro-hardness IRHD	36	55
Natural rubber (SMR 5L HC)	100	—
Natural rubber RSS1	—	100
Zinc oxide	3	5
Stearic acid	0·5	2
Nonox WSP	1	—
Barytes	60	—
Red Ochre	1	—
MBTS	1	—
Sulphur	1·5	2·5
CBS	—	0·5
Antioxidant 4010	—	1
Flectol H	—	1
SRF black	—	50
Dutrex R	—	5
Mooney viscosity ML3 120°C	47·5	52
Mooney scorch 120°C, T5 min.	36	36

The first NR compound (mix, 43; Table 9.19) which was especially designed by NRPRA as an adrenaline bottle stopper was particularly suitable for use in this type of application and it filled the mould in

8 sec. at an injection temperature of 130°C. The engineering compound containing SRF black injected at 136°C, filled the mould in 8 sec. and cured in 35 sec. at 180°C.

9.7 COMPOUNDING NR FOR INJECTION MOULDING: CONCLUSIONS

A very wide range of conventionally cured NR compounds can be injection moulded in attractively short cure times at 180–200°C depending on section thickness:

Thickness—cm.	0·16	0·28	1·0	2·1	3·8	9·8
Cure—sec.	10	15	45	60	60	360

The list of compounds studied includes gum and filled compounds having a hardness range of 40–91 IRHD and a Mooney viscosity (ML3, 120°C) range of 32–110. These figures do not in any way indicate that compounds outside these limits could not be handled quite satisfactorily. The list includes compounds containing black fillers such as MT, SRF, HAF, FEF, and MPC blacks at 50–100 phr loadings, which are satisfactory for engine mounts, commando shoe soles and solid truck tyres. The range of white fillers examined includes clay, precipitated whiting, aluminium silicate and silica.

A number of compounds containing a variety of curing systems having Mooney scorch times (T5 at 120°C), from 9–45 min. were satisfactorily moulded, but compounds with scorch times less than 9 min. could not be handled with the Daniels 45 SR machine when the barrel was 90°C and injection temperature 120–130°C.

Eight different injection machines have been successfully used with the NR/50 SRF black compound and in several machines a very wide range of moulding conditions have given satisfactory mouldings.

It may be concluded that almost any NR compound can be injection moulded and that rubber product manufacturers may safely carry out first trials with the existing compound for the job. However, it does not follow that such a compound will give the best rate of production that may be obtainable by subtle use of machine controls and careful compounding.

One of the key factors in injection moulding is obtaining the highest possible injection temperature consistent with freedom from scorch in the barrel or moulding (Table 9.3). Evidence in Table 9.12 indicates that a 1½-in. thick section can be cured quicker with a relatively safe sulphur/CBS curing system than with a similar, but boosted and therefore potentially faster curing system. The sole reason for this is that the barrel can be maintained at a higher temperature, thus giving a higher injection temperature. In addition, when injection temperatures are relatively low compared with mould temperatures, reversion of a thin section may occur before an adjacent thicker section is fully cured (Table 9.7).

Consequently, it is advantageous to choose an injection moulding accelerator with a high safety factor and good plateau curing characteristics. The high injection temperature obtained allows the minimum of time delay for mould heating and minimizes the risk of reversion

Conventional accelerators are satisfactory for injection moulding under many conditions, but even the 'delayed action' sulphenamide accelerators tend to be 'peaky' at 180–200°C. If reversion is troublesome, plateau characteristics may be improved by stepwise simultaneous reduction of sulphur levels and increase in accelerator levels (Fig. 9.35). The ultimate in this is use of 0·25–0·4 phr sulphur in conjunction with 2·1 phr Santocure MOR and 1 phr TMTD—one of the NRPRA efficient vulcanization (EV) curing systems [35, 36, 37] which will almost certainly resolve any reversion troubles.

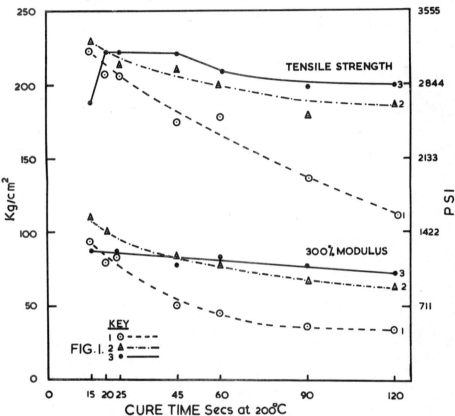

Fig. 9.35. *Improvements in tensile strength and modulus reversion by simultaneous reduction of sulphur level.*
Key: (1) sulphur, 2·5 phr; Santocure, 0·5 phr; (2) sulphur, 2·0 phr; Santocure, 1·2 phr; TMTD, 0·3 phr; retarder, 1·0 phr; (3) sulphur, 0·25 phr; Santocure MOR, 2·1 phr; TMTD, 1·0 phr.
Injection moulding of 0·28 cm. thick sheet with Peco 21TS machine and mixes as Table 9.10.

In view of the extra cost of EV curing systems and possible reduction of flex life under certain conditions of use where dynamic strains pass through zero strain, stepwise reduction of sulphur is advised until the reversion problem is beaten. In any application where ageing is important flex life of EV compounds is superior to that of conventional compounds.
Apart from the considerations already mentioned no special compounding

innovations have been found necessary to date. As the properties of a finished product vary with compound formulation it is better to adjust machine controls to obtain optimum production rates before making extensive compounding changes.

In view of the importance of filling a mould in as short a time as possible (e.g. 1 or 2 sec.) to prevent scorch one cannot over stress the importance of obtaining the highest possible injection temperatures and flow rates consistent with freedom from scorch. The value of nozzle diameter control is also important.

There may be a temptation to adjust compound viscosity by using plasticizers, peptizers or extenders to obtain faster mould-filling times, but it is better to increase barrel temperatures and/or screw speeds and Mooney scorch times than to resort to artificial softeners or extenders.

In experiments where as many injection moulding machine factors as possible were held constant, addition of softeners gave only a small reduction in mould-filling time (injection time), but at the cost of lower injection temperatures and markedly longer curing times (Table 9.8). This effect is more exaggerated as the section to be cured increases in thickness, especially if temperatures are 140°C or lower (Table 9.13).

For very soft stocks, it may be advantageous to compound with a processing aid [41] such as PA80 (25 PA80, 75 RSS). This keeps the mix stiff enough to provide resistance to the ram, thus deriving the maximum benefit for the stock as it is injected through the narrow constriction into the mould. As much as 50% PA80 may cause poor feeding if the mix 'breaks short' when meeting the screw.

Izod and Morris [29] found that under similar conditions of barrel temperature, nozzle diameter and injection pressure, heat build-up of an NR compound through the nozzle was 35°C compared with only 10°C for a similar cis-polyisoprene compound.

The useful heat build-up of NR is seen as an advantage where production rate depends on injection temperature and it can be seen as a safety factor when one considers that, for the same injection temperature, an NR compound with a high nozzle heat build-up may remain in the barrel with less risk of scorch than a compound based on a polymer which affords a lower heat build-up.

NR mixes containing all the fillers mentioned earlier afford a useful heat build-up during passage through the screw and nozzle. HAF black generates more heat build up than SRF black. A mix containing precipitated calcium carbonate gives a good heat build-up, similar to that given by SRF blacks, while precipitated aluminium silicate gives very high values and was the scorchiest type of mix to handle.

By contrast, china clay mixes are cool running. Since a highly loaded clay mix injects more coolly than one with a lower loading it appears that (if product properties permit it) injection temperature can be influenced by choice or level of filler.

Thin films of injection moulding flash, especially from white-filled or poorly protected mixes, tend to become sticky at 200°C. Polymerized 2,2,4-trimethyl 1,2 dihydroquinoline may be recommended for heat resistance [41, 42], and the best protection against stickiness yet observed on a purely subjective basis was obtained by using this antioxidant in conjunction with 1 phr TMTD in the curing system backed up by 1 phr of Antioxidant 4010, in a compound designed for dynamic usage.

This work forms part of the research programme of the Natural Rubber Producers' Research Association.

ACKNOWLEDGEMENTS

We acknowledge with thanks the cooperation of the Rubber and Plastics Research Association in the carrying out of the injection moulding with the Daniels 45 SR machine in their laboratories. We also thank T. H. and J. Daniels Ltd for permitting the machine to be placed at our disposal.

We also gratefully acknowledge the help of the staffs of Peco Machinery Ltd during injection moulding with the Peco 21TS injection moulding machine, David Bridge and Co. during injection moulding with the Farrel-Bridge Model R60-350 rubber injection moulding machine, Desma during injection moulding with the four-station Desma 905 machine, Stübbe during injection moulding with the Stubbe S150/235 machine and Seidl during injection moulding with the Seidl SPA 1 BX-A machine and we thank them for placing the moulding machines and moulds at our disposal.

REFERENCES

1   *Rubber World,* July, 1963, **148**, 29.
2   Bament, J. C., Paper to Swedish Assn. of Rubber Technology, June 3–4, 1965.
3   *Plastics and Rubber Weekly,* June 4, 1965.
4   *Rubber Journal,* 1965, **147**, 7, 72.
5   Dale, B., Ibid., 1964, **146**, 1, 50.
6   Krol, L. H., Verkerk, G. and van der Grijn, J., *Rubber and Plastics Age,* 1963, **44**, 3, 284.
7   Hollis, E. W., *Rubber Age,* 1961, **90**, 2, 261.
8   *Rubber Journal,* 1965, **147**, 11, 64.
9   Kleine-Albers, A., and Franck, A., *Rubber and Plastics Age,* 1963, **44**, 5, 515.
10   Jurgeleit, H. F., *Rubber Age,* 1962, **90**, 5, 763.
11   Lane, R. G., Ibid., 1963, **93**. 6, 915.
12   Gregory, C. H., *Rubber Journal,* 1964, **146**, 8, 46.
13   Anon., *Rubber Journal,* 1967, **149**, 10, 60.
14   Cumming, A. P. C., *Rubber Journal,* 1967, **149**, 12, 33.
15   *Rubber and Plastics Weekly,* June 22, 1963, 900.
16   Booth, D. A., Paper to Division of Rubber Chemistry, ACS, Oct. 21, 1965.
17   Wheelans, M. A., *Rubber World,* 1967, **156**, 5, 72.
18   *India Rubber Journal,* 1944, **107**, 529.
19   Vanderbilt, R. T., Co., Inc., Injection Moulding and High Temperature Press Curing of Natural and Synthetic Rubbers, March 15, 1946.
20   Hendrick, J. V. and Fraser, D. F., *Rubber Age,* 1944, **56**, 277.
21   Fraser, D. F., *India Rubber World,* 1948, **118**, 357.
22   *Rubber Age,* 1946, **59**, 488.

23   Seidl, T., lecture at Symposium on The Injection Moulding of
      Rubbers, National College of Rubber Technology, London,
      June 22, 1964.
24   Harrison, P. F. and Seklecki, T., paper to Swedish Assn. of
      Rubber Technology, May 1966
25   Daniels, D. L., *Rubber and Plastics Age*, 1963, **44**, 11, 1333.
26   Wheelans, M. A., *Rubber Developments*, 1965, **18**, 4, 133.
27   Wheelans, M. A., *Rubber Journal*, 1966, **148**, 12, 26.
28   Wheelans, M. A., *Rubber Journal*, 1967, **149**, 1, 59.
29   Izod, D. A. W. and Morris, W *Rubber and Plastics Age*, 1965,
      46, 2, 167.
30   House literature, Desma-werke, Achim Bei Bremen, Germany.
31   House literature, Maschinenfabrik Albert Stübbe, Vlotho, Germany.
32   Simon, E., lecture at Rubber Injection Moulding Exhibition,
      Delft, Netherlands, Feb. 26, 1965.
33   *Rubber and Plastics Weekly*, April 24, 1967
34   Izod, D. A. W., private communication.
35   NR Technical Information Sheets No. 78, 79, 86 and 94,
      published by NRPRA.
36   Skinner, T. D. and Watson, A. A., *Rubber Age*, 1967, **99**, 11, 76.
37   Skinner, T. D., Russell, R. M. and Watson, A. A., 1967, **99**,
      12, 69.
38   Dorko, Z. J., Timar, J. and Walker, J., *Rubber World*, July, 1963,
      148, 31.
39   Böckmann, A., *Kautschuk u. Gummi., Kunststoffe*, 1965, **18**,
      11, 737.
40   Jäkel, H., *Kautschuk u. Gummi, Kunststoffe*, 1964, **17**, 4, 194.
41   Baker, H. C. and Stokes, S. C., Proceedings of the Natural Rubber
      Research Conference, Kuala Lumpur 1960.
42   Fletcher, W. P. and Fogg, S. G., *Rubber Age*, 1959, **84**, 4, 632.
43   Bell, C. L. M., Cain, M. E., Elliott, D. J. and Saville, B.,
      Kautschuk u. Gummi, Kunststoffe, 1966, **19**, 3, 133.

DISCUSSION

W. J. LEAR *(Thiokol Chemicals)*: Could Dr Wheelans shed some light
on the possible use of peroxide-cured natural rubber bearing in mind
the problem of reversion and also components for good ageing properties?
ANSWER: Peroxide-cured natural rubber compounds certainly have very
good reversion resistance, ageing resistance and compression set.
Unfortunately natural rubber with 3·0 phr dicumyl peroxide, one of
the most popular peroxides for curing rubber, has a strong tendency to
scorch (Mooney scorch time 120°C, L + 5, 3-4 min.) and is liable to
produce sticky flash which might be expected to cause mould fouling.
For these reasons NRPRA has not yet tried injection moulding with
peroxide curing systems.

H. A. COOK *(Esso Chemicals Ltd)*: Would Dr Wheelans care to elaborate
on the type and level of 'amine' as recommended in the principles of EV
system development?
A: The type and level of amine concerned is fully dealt with by Skinner
and Watson (*Rubber Age*, 1967, **99**, 12, 69). If MBT (or MBTS) is the
accelerator concerned a molecular quantity of amine, equivalent to the

MBT is required to solubilize the zinc salt of MBT

$$X - S - Zn - S - X \quad \text{where X is}$$

Sulphenamide accelerators such as CBS may be considered to consist of an equimolecular proportion of amine and MBT.

W. S. PENN *(Borough Polytechnic)*: Would the use of blends of NR and SBR not overcome the problem of reversion?

A: The use of SBR (and BR) in blends with NR does help to improve reversion resistance where conventional curing systems are used.
The improvement is not so marked as that produced by changing the curing system to an EV system. One should also consider the properties of the vulcanized product and, where resilience is the desirable property, this is likely to be spoiled by the presence of a more hysteretic polymer. Since natural rubber is an excellent polymer for injection moulding purposes there is no particular need to introduce other polymers, especially if there is a danger of reducing the general high level of properties associated with NR.

D. A. BOOTH *(Esso Research SA)*: Where you have shown differences in certain physical properties between injection and compression moulded samples, have you been able to confirm differences in state of cure (crosslink density) by techniques such as volume swell?

A: Comparisons between injection and compression moulded samples have been made by conventional physical testing procedures and no measurements have yet been made by swelling techniques.
Measurements have so far been confined to modulus, tensile strength, tear strength, resilience, hardness and ageing tests.

CHAPTER 10

Injection moulding of general purpose synthetic polymers

L. M. GLANVILLE
The International Synthetic Rubber Company Ltd

10.1 INTRODUCTION

DURING the past five years, the use of the injection moulding process has gained ever increasing importance to the rubber industries of Western Europe and the U.S.A. One outstanding cause of this has been the steady rise in labour costs in these parts of the world.

The idea of injection moulding dates back to the late 1930s when attempts were made in the U.S.A., to design equipment for this purpose. The fact that nothing was heard of this development until comparatively recently, suggests that, for some reason or other, the project was discontinued.

Much of the recent progress of injection moulding in the rubber industry arises from the wide experience gained by the machinery manufacturers engaged in supplying equipment for procedures of thermoplastic mouldings. This experience is now slowly being extended to the rubber industry particularly in view of the claims which are being made (cf. Chapter 9) for the advantages of injection moulding over compression and even over transfer moulding.

10.2 SYNTHETIC POLYMERS AT HIGH TEMPERATURES

The polymers considered in this chapter are as follows:

1. Styrene butadiene (SBR) polymers (regardless of their bound styrene level or if oil extended or not).
2. cis-Polybutadiene (BR)
3. Natural rubber (NR) and cis-polyisoprene (IR).

When rubber manufacturers first began to give serious consideration to injection moulding much was written about the suitability of various polymers for this process and it would be worth while to consider briefly the behaviour of these polymers when they are vulcanized at high temperatures.

First, one feature which general purpose polymers have in common is that as the temperature of vulcanization is increased, the tensile strength values obtained from each type of polymer are significantly reduced by varying magnitudes.

Figure 10.1 illustrates this effect. All the values shown were obtained by determining the optimum cure time on a Shawbury curometer and the same batch of compounded polymer was used for obtaining the tensile values for each of the three curing temperatures quoted.

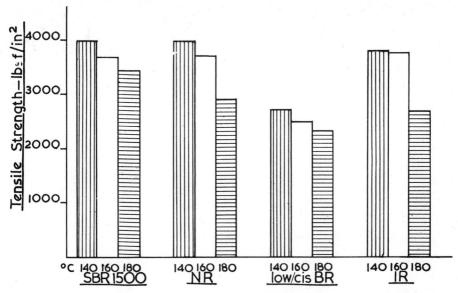

Fig. 10.1. Effect of increased cure temperature upon tensile strength.

This data indicates that at vulcanizing temperatures above 160°C, there is a significantly more rapid loss in tensile strength for NR and cis-IR than for SBR and BR polymers.

Under these conditions NR and cis-IR appear to be more susceptible to reversion as indicated by a loss in tensile strength. Under the same circumstances, SBR tends to form oxidative crosslinks which offset this reversion effect. This is indicated by a gradual resinification effect, i.e. there is a tendency for compounds based on SBR to stiffen progressively and harden without the loss being so marked in physical properties. This particular characteristic can be varied to some extent by the type of antioxidant subsequently used by the manufacturer.

It should be noted that low cis-BR differs from SBR in as much as the oxidative crosslinking that occurs is to a much lesser degree, and reversion does not predominate to such a degree as with NR and IR. This may be explained by the normal existence of 8–9% vinyl groups in low cis-BR which may tend to form oxidative crosslinks during ageing. Also the use of a lithium butyl catalyst will result in the absence of metallic residue which could act as depolymerizing agents.

A practical example of where the reversion effect can cause problems during the injection moulding process, is when the thin feathery flash which tends to form around cavities becomes sticky if it is not removed after each curing cycle. This could then find its way into a cavity and contaminate subsequent mouldings and necessitate the cleaning of the cavity.

If NR or cis-IR is blended with SBR, or low cis-BR, then the risk of mould fouling because of reversion is greatly reduced. Another advantages that often arises from using low cis-BR in all types of compound regardless of the moulding technique, is that a slower rate of build-up on the mould surfaces is often obtained.

10.3 SBR AND BR POLYMERS IN INJECTION MOULDING

When the rubber industry first began to take positive steps to adopt the injection moulding process much was being written on the suitability of various types of polymers for this purpose. Therefore the International Synthetic Rubber Co. sought to obtain some basic knowledge on how its polymers, namely SBR and BR, would perform when used under these conditions. A programme of work was carried out by ISR in collaboration with rubber injection moulding machine manufacturers T. H. and J. Daniels of Stroud.

The sole object was to assess the suitability of Intol SBR and blends of Intol SBR and Intene BR as polymers bases for compounds processed under injection moulding conditions. In this aspect of the study an evaluation of the level of physical properties obtained from such compounds using both compression and injection moulding techniques was carried out.

A range of compounds was selected without any consideration to their probable suitability for the injection moulding process and comprise the following:

Compound	Type
1	Direct vulcanization shoe soling compound based on 50/50 SBR/BR blend.
2	Direct vulcanization shoe soling compound based on 100 SBR.
3	Tyre tread type compound based on 100 oe SBR.
4	First quality mechanical compound based on 60/40 SBR/NR blends.
5	Second quality mechanical compound based on 100 oe SBR.
6	Grade II resin soling compound based on 100 oe SBR.
7	Translucent shoe soling compound based on 60/40 oe SBR/BR blend.

In order to obtain a realistic comparison it was decided that the physical properties obtained by the injection moulding process would be compared with those obtained by the compression moulded method. Conventional curing times and temperatures were used for the latter method rather than to attempt to prepare compression moulded test pieces at the short cure times and very high temperatures used in injection moulding. Therefore factory curing conditions were simulated as near as possible.

Another feature of this investigation was that these compounds provided a wide range of Mooney viscosities (ML4 at 100°C) —40 to 75 and Mooney scorch times ranging from 4 to 11½ min.

Tables 10.1–10.6 The formulations include in addition to variations in base polymer, a range of reinforcing and non reinforcing white fillers and carbon black, high styrene resin, differing accelerator systems and sulphur levels and a sulphur donor vulcanizing system. These compounds were chosen to give varying compound and vulcanizing characteristics. The compounds were conventional and not specifically designed for injection moulding purposes.

All seven compounds were subjected to moulding trials in the

Table 10.1 DV soling compounds

	1	2
Intol 1509	50	100
Intene 35NF	50	—
FEF black	30	30
Aluminium silicate	50	50
China clay	20	20
Zinc oxide	5	5
Stearic acid	2	2
Coumarone-indene resin	2·5	2·5
Naphthenic oil	20	20
Diethylene glycol	1·5	3
CBS	1	1
MBTS	1	1
TMT	0·5	0·5
Sulphur	3·5	3·5

Table 10.2 Tyre tread compound

Intol 1712	100
HAF black	50
Zinc oxide	4
Stearic acid	1
Aromatic oil	5
Antioxidant	1
MBT	0·75
DPG	0·75
Sulphur	2

Table 10.3 First quality mechanical compound

Intol 1500	30
Intol 1712	30
Intene 55NF	40
FEF black	55
Zinc oxide	3
Stearic acid	1·5
Aromatic oil	20
Paraffin wax	1
PBN	1
4,4'–Dithiodimorpholine	0·5
TMT	3

Table 10.4 Second quality mechanical compound

Intol 1714	100
China clay	175
Activated calcium carbonate	25
Zinc oxide	5
Stearic acid	1·5
Coumarone-indene resin	7·5
PBN	1·5
Diethylene glycol	2
MBT	2
DPG	0·5
Sulphur	1·75

Table 10.5 Resin soling compound

Intol 1778	100
85/15 Styrene/butadiene resin	30
Aluminium silicate	75
China clay	20
Titanium dioxide	3
Zinc oxide	5
Stearic acid	2
Coumarone-indene resin	2·5
Diethylene glycol	3
Antioxidant	0·5
MBT	1
TMT	0·5
Sulphur	3·5

Table 10.6 Translucent soling compound

Intol 1778	60
Intene 35NF	40
Silica	50
Active zinc oxide	1·5
Stearic acid	1·5
Naphthenic oil	10
Diethylene glycol	3
Triethanolamine	1
Activated thiazole	1·75
ZDC	0·75
Sulphur	1·75

Daniels/Herbert Edgwick 45 SR rubber injection moulding machine. The mould used in these trials gave a product resembling a squat flower pot having a base diameter of 3 in. The wall thickness of the mould was 0·08 in. and the single injection part was situated in the centre of the base of the moulding (see similar moulding Fig. 9.2). This moulding proved particularly suitable for the preparation of physical test pieces, these test pieces being cut from the wall of the moulding in a direction parallel to the base. It was established that test pieces cut from the wall of the moulding at right angles to the base showed no significant difference in values obtained. In view of this, it can be assumed that anisotropy is not a significant factor in SBR or SBR/BR compounds when the injection moulding process is used.

Conventional methods were used for the preparation of compression or moulded test samples, these samples being vulcanized to optimum for tensile strength at 144° or 150°C. Both the compression and injection moulded vulcanizates were prepared from the same batches of compound.

EDGWICK 45 SR INJECTION MOULDING MACHINE CONDITIONS

The machine conditions given in Table 10.7 were used in the preparation of the injection moulded test pieces. The compounds were fed to the injection machine either in strip form or in a chipped condition from a Masson cutter.

Table 10.7 Edgwick 45 SR injection moulding machine conditions

Injection pressure psi	8 200–10 930
Hold-on pressure psi	5 470
Injection time sec.	5–8
Cycle time min.	½–4
Barrel temperature °C	71
Mould temperature °C	177
Screw speed rpm	50–75
Injection nozzle diameter in.	$\frac{3}{32}$

The cycle times referred to include the few seconds required for injection of the compound into the mould. Actual cure times will therefore be the cycle time minus the time required for injection of the compound.

All seven compounds previously described proved capable of being moulded without difficulty using the Daniels/Herbert Edgwick 45 SR injection moulding machine. In all instances the moulding possessed excellent definition with very little mould flash. There was no evidence of moulding defects including trapped air blisters and a complete absence of flow lines within the vulcanized compound.

10.4 DISCUSSION OF RESULTS

The findings of this study may now be summarized. The wide degree of latitude with regard to compound Mooney and Mooney scorch values is indicated in compounds based on SBR and blends of SBR and BR. For example, a compound with a Mooney viscosity value of 75 and a scorch time of just under 4 min. (compound 7) processed without difficulty giving excellent mouldings without any form of blemish, and improved tensile strength.

The following illustrations give some indication of the effect of the injection moulding process on the physical properties of all seven compounds evaluated and represent the optimum values obtained in each instance:

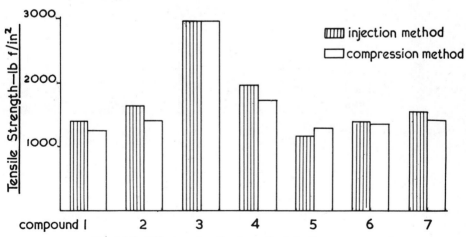

Fig. 10.2. Effect on tensile strength—injection v compression.

It is evident from Fig. 10.2 that the tensile strength characteristics of
SBR and SBR/BR based compounds are slightly improved by the
injection moulding process. This is also regardless of the fact that a
mould temperature of 177°C was used whereas temperatures of 144°
or 150°C were used when obtaining the compression moulding physical
test data. This improvement in tensile strength is most likely the result
of improved dispersion of reinforcing fillers caused by the high rate of
shear to which the compound is subjected during injection.

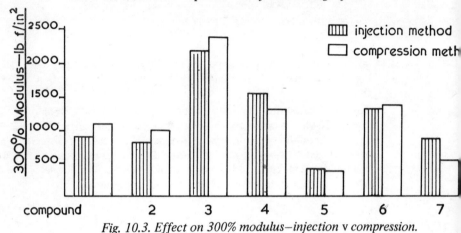

Fig. 10.3. Effect on 300% modulus—injection v compression.

From Fig. 10.3 there appears to be no obvious pattern to the effect of
injection moulding on the modulus values obtained.

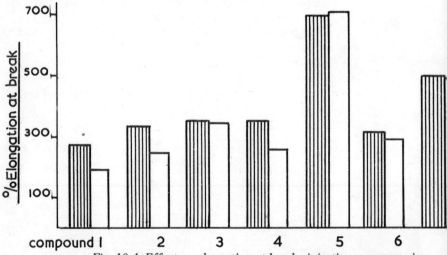

Fig. 10.4. Effect on elongation at break—injection v compression.

From Fig. 10.4 it may be seen that the elongation at break values given
by the injection moulding process tend to be higher than those from
compression moulding.

It may be seen from Fig. 10.5 that the hardness values resulting from
injection moulding are usually slightly lower than for compression
moulding.

134

PLATE 15

Fig. 9.34. The 90 cavity blood test phial stopper mould illustrating the single injection point and the complex runner and gate system for filling each cavity. The gates can be seen to increase in size the further they are from the injection point.
(Courtesy of NRPRA).

Fig. 12.20. Three-insert runners and gate mould (moving half).

Fig. 12.21. Cast from Fig. 12.20 (above).

PLATE 16

F⊢ 1 INCH ⊣

Fig. 12.23. Feedstock. (1) Small irregular granules from Masson cutter. (2) ¼in. cube. (3) ½ x ½ x ¼in. granule. (4) 1 x 1 x ¼in. size.

Fig. 12.32. Multiple test piece injection mould (moving half).

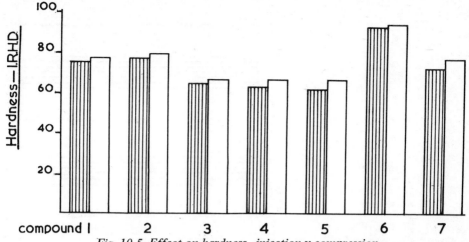

Fig. 10.5. Effect on hardness—injection v compression.

A suggested reason for these last two effects is that the SBR polymers undergo a limited degree of physical degradation when subjected to high shearing forces at temperatures in the order of 120°C.

The retention of physical properties after seven days ageing at 70°C in air of the injection moulded vulcanizates included in this study is excellent and compares very favourably with the aged values obtained from compression included vulcanizates (Figs 10.6 and 10.7).

Although there is no illustration for this point it has been found that the high degree of retention of physical properties is obtained over a wide range of injection cycle times.

Little limitation would appear to exist on the choice of compounding ingredients for the design of compounds for injection moulding purposes based on SBR or SBR/BR blends.

The range of compounds used in this study includes compounds based

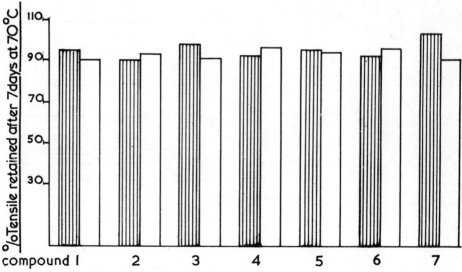

Fig. 10.6. Effect of ageing on tensile strength—injection v compression.

135

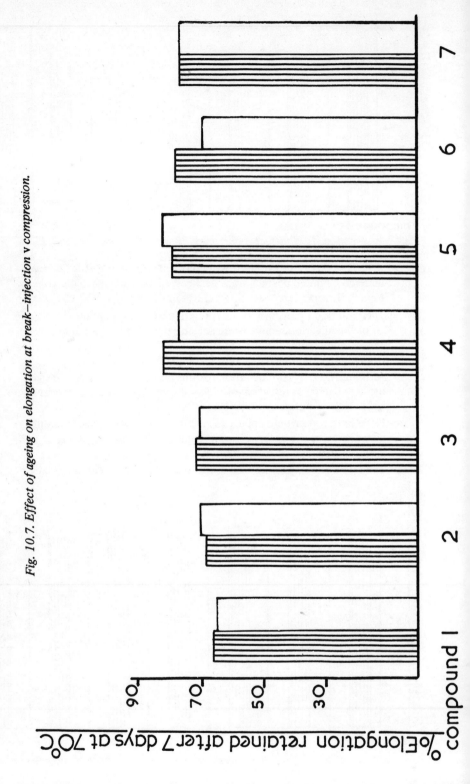

Fig. 10.7. Effect of ageing on elongation at break—injection v compression.

on a range of polymers, compounds reinforced with carbon black, fine particle silica and varying mixtures of carbon black, aluminium silicate china clay and surface treated whiting. Variations in vulcanizing systems and a sulphur donor curing system have also been successfully employed.

Our conclusions from this study are that SBR and low cis-BR/SBR blends are ideal for injection moulding purposes. No adverse effect upon physical properties, when, compared with normal compression moulding methods is likely to occur. Both SBR and low cis-BR can be expected to enhance the heat stability of NR or cis-IR when used in blends for high temperature curing applications.

DISCUSSION

W. S. PENN *(Borough Polytechnic)*: Can you please comment more on moulding fouling?

ANSWER: It has been frequently reported by our sales representatives that customers have found that moulding compounds, partly based on Intene, have given low rates of undesirable build-up of carbon deposits on mould surfaces during prolonged production runs.

This build-up problem usually becomes more pronounced as moulding temperatures are increased.

Two suggested reasons why Intene behaves in this way are:
1 Intene is highly resistant to reversion and does not deploymerize so markedly as NR or poly*iso*prene at high temperatures.
2 Polybutadiene is highly compatible with most petroleum-based products and would greatly lessen the risk of their oxidation on the surface of the rubber compound by retarding any tendency for them to exude to the surface at high temperatures.

R. MCEWAN *(Phillips Petroleum U.K. Ltd)*: I agree that solution polymerized rubbers give better physical properties than NR at high temperatures. Has the speaker compared the properties of solution polymers *v.* NR using EV system for curing both?

A: As yet, we have not had the opportunity to evaluate EV systems and our knowledge is strictly confined to what we have read in various technical publications.

W. S. PENN: Would the use of blends of NR and SBR not overcome the problem of reversion?

A: Yes they would to a great extent, but some people have expressed the opinion that the optimum physical properties cannot be obtained from the respective polymers in such a blend. Certainly there is a difference in cure rates which can eventually have an adverse effect on the physical properties after ageing.

Polybutadiene has the advantage of having curing characteristics which are very similar to those of the best grades of natural rubber. This could be to the benefit of the aged characteristics.

H. M. GARDNER *(T. H. and J. Daniels Ltd)*: The results illustrated are for standard 'compression' compounds, used under injection conditions. What improvements could be expected if compounds were specifically tailored for injection?

A: The improvements from 'tailor-made' injection moulding compounds would be more of an indirect nature. The ability is required to design curing systems which will provide an adequate level of processing safety, and storage stability, before the compounds are vulcanized. At this stage, the curing systems have to function extremely efficiently in curing cycles which may be specified in seconds for curing temperatures above 165°C. This aspect can be further complicated by the thickness of the section to be vulcanized. This would be the main 'improvement' that could be expected.

We would like to emphasize that, apart from such details as these, the necessity for special compounding techniques does not arise.

CHAPTER 11

Injection moulding of nitrile rubber compounds

B. DALE, B.Sc.(Oxon), A.R.I.C., A.B.I.M.
B.P. Chemicals (U.K.) Limited

11.1 INTRODUCTION

THE injection moulding of vulcanized rubber compounds has been undertaken for limited and specialized purposes for many years. Recently a number of injection moulding machines, of both the ram and screw type, have become commercially available, from France, Austria, Japan, America, Germany and this country, and the rubber industry's general interest in them is very great. There is no doubt that the more economic processing and greater productivity of these machines will be an important factor in the future development of that sector of the rubber industry devoted to high volume repetitive production of custom mouldings, rubber to metal bonded parts and D/V shoe soling.

The advantages of injection moulding have been listed many times. To some extent the advantages of injection moulding can be matched by modern high speed compression presses, which, operating at platen closure speeds of 12 in. per sec. and fitted with electrically heated platens reaching temperatures of 200°C, can approach the output rates of injection moulding. The finer points of the economics of the press versus injection methods can only be judged by reference to the size, thickness, and shape of the component and the degree to which the feed of raw material to the machine and the removal of finished parts, can be rendered easy or automatic.

The investment in injection equipment provides savings in unit cost. This generalization can be broken down into more detail. For example, mould costs are higher, but on the other hand fewer mould cavities are required for any given job. It is to be expected that contracts can be completed more quickly, giving greater flexibility in shop loading. There is no doubt that during the last year or two the interest and attention given to studies of faster production cycles, by the injection moulding process, has had a tremendous effect in speeding up compression moulding practice, and both techniques have benefited by the competition.

Single station injection moulding machines are available from roughly 4 oz. capacity at £3 500 up to 60 oz. at £23 000. Types are also available with multiple mould stations. Each type has its own place in the industry, but for general purpose work the screw feed type seems to have the advantage of kneading the compound together and bringing it to a steady controlled temperature ready for injection. With the screw type machine higher effective injection pressures are developed owing to the lower friction on the walls of the cylinder. With cycle times of less than 1 min. the multi-station machines would appear particularly advantageous in the shoe sole direct vulcanization process when uppers are fitted in to the mould tool and where a little more time has to be allowed for loading and unloading bonded parts. Very long runs, of automotive parts for example, are also economic for multi-station machines. Special purpose machines have now been designed and put into operation for making one type of component only. When one considers the special purpose applications of nitrile rubber, and its small tonnage compared with general purpose rubbers, the significance of building a special machine to make parts with it is a powerful justification for the advanced injection moulding technique.

The processing of rubber by injection moulding appears now to be so highly developed that very little difficulty will be experienced in starting up and running. The emphasis on compounding technology will be in the refinement of the vulcanization system and the viscosity of the mix so as to achieve higher and higher outputs.

11.2 TYPES OF COMPOUNDS INVESTIGATED

There is nothing particularly novel about the injection moulding of rubber compounds and nitriles are no exception in that they have been successfully processed in this way for many years. However, in view of the current interest it was decided to carry out work, both in the technical service laboratories of BP Plastics and in conjunction with machine suppliers. Rheological studies on unvulcanized compounds and a programme of compression moulding at normal and higher temperatures was carried out in conjunction with machine trials at the works of T. H. and J. Daniels Ltd of Stroud.

Three compounds were selected for examination (Table 11.1)

Compound No.

57082/1563	High nitrile content, hot rubber	Breon nitrile 1001
57082/1562	Medium high nitrile content, cold rubber	Breon nitrile 1042
57082/1565	pvc/nitrile blend, ratio 55 : 45	Breon Polyblend 503

The first two recipes used in this work were chosen to represent normal plasticized black filled nitrile compounds which can be extruded and vulcanized on the more conventional present day equipment. These are typical of the basic formulations which with simple modifications can be used for many of the moulded articles which are at present compression moulded for the motor and aircraft industries.

The curing system is typical of the type at present used for the fast production of compression mouldings at high temperatures (180-200°C) being characterized by a fairly long Mooney scorch period at 120°C with rapid development of physical properties at the curing temperatures.

The formulations differ only in the base polymer; 57082/1563 is based on Breon 1001, a high nitrile type produced by hot polymerization and compound 57082/1562 on Breon 1042, an easy processing medium high nitrile rubber manufactured by the 'cold' technique. Compound 57082/1565 is based on Breon Polyblend 503, a 55/45 blend of a

140

medium nitrile and pvc. No filler was used in this compound so as to give the lowest possible Mooney viscosity Otherwise it is similar to the previous compounds.

It was intended that these three compounds should show up differences in processing of the basic polymers rather than that they should be necessarily representative of actual compounds to be employed by processors for particular applications.

PROPERTIES OF POLYMERS, UNVULCANIZED AND VULCANIZED STOCKS (TABLE 11.1)

It is usually possible to obtain an idea of processing properties, when compression moulding, extruding or calendering, from laboratory data such as Mooney plasticity and scorch time. This is some help in judging injection moulding processing quality, but not necessarily the whole story.

Injection times are greater with high Mooney stocks, but this is not always so important or significant as it might at first appear since very often the compounds concerned can be injected quickly enough for economic processing; or alternatively the injection pressure can be raised, or the injection nozzle opened up, so as to speed up the cycle.

This is particularly true with the screw type of machine as there is greater mechanical efficiency in reserve to overcome the resistance of the stock. Another variable is the stock temperature in the injection cylinder. At 70°C, which is a quite reasonable temperature at which to hold a nitrile compound for half an hour, stocks are fluid and injection times are low, i.e. 5–15 sec. typically.

The properties of the raw polymers; stocks with lubricant, carbon black filler and plasticizer; and complete compounds including the vulcanization systems are given in Table 11.1.

Points worth noting are (a) the wide variation in 120°C Mooney ML_4 viscosity of the three compounds from 63 at one extreme to 24 at the other and (b) the wide variation in Mooney scorch at 120°C, from 8 min. to 42 min. Fig. 11.1 puts the Mooney scorch values in graphical form. The vulcanized properties of the three stocks are reasonably close and this is due to the selection of 'simple' compounds without extenders or specific ingredients employed to adjust properties to meet marginal requirements (Table 11.1).

The variations in plasticity and cure, however, cover most of the range likely to be employed by actual fabricators of finished parts.

11.3 MACHINE DETAILS

The machine used was the Edgwick 45 SR and is based on the standard Edgwick in-line screw preplasticizing plastics injection moulding machine. The significant modifications which have been introduced to make it convenient for the injection moulding of rubber are to fit it with a thermostatically controlled water jacket for heating, cooling (including emergency cooling of the barrel) and thermostatic controls for the heater bands on the two halves of the mould.

The machine is neat in appearance and well designed with all controls

Table 11.1
Table of compounds

	57082/1563	57082/1562	57082/1565
Breon nitrile 10011	100	–	–
Breon nitrile 1042	–	100	–
Regal SRF	65	65	–
Bisoflex 791	10	10	10
Zinc oxide	5	5	5
Stearic acid	0·75	0·75	–
MC Sulphur	0·5	0·5	0·5
TMT	2	2	2
MBTS	2	2	2
Breon Polyblend 503	–	–	100
Mark 33	–	–	2

Properties of unvulcanized compounds

	57082/1563	57082/1562	57082/1565
Polymer base	Nitrile 1001	Nitrile 1042	Polyblend 503
ML_4 polymer, at 100°C	99	68	72
ML_4 compound without curing system, at 100°C	99	70	36
ML_4 compound with full curing system, at 120°C	63	39	24
Mooney scorch of compound at 120°C with curing system (5 pt rise time on minimum)	8 min.	26 min.	42 min.
Specific gravity	1·18	1·19	1·18

Properties of vulcanized compounds

	57082/1563	57082/1562	57082/1565
Compression moulded 15 min. at 153°C			
Tensile strength lb/in^2	2 600	2 320	2 060
Elongation at break %	305	415	320
Modulus at 100% lb/in^2	630	357	595
Modulus at 200% lb/in^2	1 700	996	1 030
Modulus at 300% lb/in^2	2 540	1 687	1 810
Tear strength, lb. BS 903, room temp.	30·3	36·8	27·0
150°C	10·9	12·8	2·8
Hardness °BS	69	63	68

centralized for ease of operation. Screw rotation is effected by means of a hydraulic motor driving through a mechanical gear box. This ensures that maximum torque is maintained even at the low speeds that would be required for very stiff stocks.

The principle to be followed is to run at as high rpm as possible without scorching. Eighty rpm is generally found to be a convenient setting, but this can be varied within wide limits if required.

The throat design requires to be modified for rubber strip feeding but it has been found satisfactory to employ granulated rubber compound with a vibrating slope feed interconnected automatically with the cycling controls of the main machine. Granules are produced in a small Masson cutter from strip rubber which has been given a preliminary dusting with zinc stearate (Fig. 11.2).

There are dangers inherent in the use of such dusting powders in safeguarding against moulding faults in normal compression moulding practice, but there was no difficulty with the granulated feed to this type of screw preplasticizing injection cylinder and all mouldings produced were sound internally with no signs of poor joining between flow lines. The injection screw is the general purpose thermoplastic screw fitted with a ball type non-return valve on the screw tip. It is stressed that with the three types of compound tested there were no signs of stagnation or vulcanization in the non-return valve mechanism. The cylinder nozzle has a parallel land. For some compounds it would

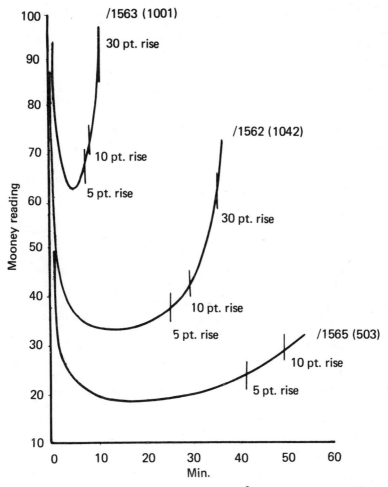

Fig. 11.1. Mooney scorch curves at 120°C for injection moulding c compounds 57082/1562 (1042), 57082/1563 (1001) and 57082/1565 (503).

be easier to adjust the cycle to ensure 100% removal of the sprue with the moulding if the land had a reverse taper, as under some conditions vulcanized rubber can be left in the nozzle to come out into the next moulding. The manufacturers have since modified the nozzle, incorporating reverse taper. This difficulty only arose with curing times in excess of those required for full vulcanization so that it is not to be expected with the normal very short cycle times being reached in production.

During the trials mould opening and ejection pin operation was manually selected, but the machine can be set up with automatic platen operation

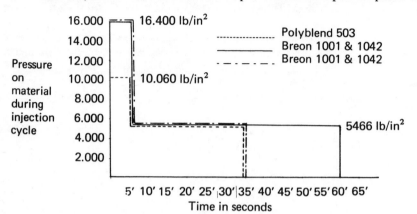

Fig. 11.2. Typical pressure/time curves for cycle times of 1 min. and over.

and automatic moulding ejection by mechanical pins or compressed air, etc according to the type of part to be produced.

It is worth noting here that the machine can be supplied with an alternative cylinder so as to have the facility to change over from rubber to plastics processing in a matter of a few hours.

The moulding produced was a flowerpot shape with 0·08 in. wall and 0·1 in. bottom, sprue centrally in base. The mould was fitted with band heaters for each half (Figs 11.3 and 11.4).

Throughout the trials the barrel temperatures was controlled at 70°C and the screw revolutions at 81 rpm. The temperature of the freely ejected stock was 100°C by needle pyrometer. The injection pressure was around 16 000 lb/in² on the nitrile stocks during injection and the first stages of vulcanization, and 10 000 lb/in² on the nitrile/pvc stock. This drops to a holding pressure of about 5 000 lb/in² during the holding part of the remainder of the vulcanization cycle, and then the screw goes back to accept a further shot of 2½ oz. of compound granules. Pressure is maintained in the mould at this stage by virtue of the resistance of the cured sprue. See graph of pressure/vulcanization cycle (Fig. 11.5).

11.4 DETAILS OF TRIALS

Compound 57082/1562–Breon nitrile 1042
(Tables 11.4, 11.5, 11.6, 11.7, 11.8).

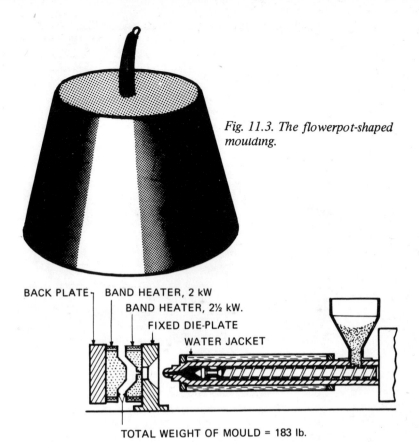

Fig. 11.3. The flowerpot-shaped moulding.

BACK PLATE
BAND HEATER, 2 kW
BAND HEATER, 2½ kW.
FIXED DIE-PLATE
WATER JACKET

TOTAL WEIGHT OF MOULD = 183 lb.

Fig. 11.4. The injection system, mould tool and band heaters

Thirty-six injection mouldings were produced and examined fully for physical properties. Mould temperatures were 177°C, 188°C, 193°C and 205°C. Cycle times including injection and vulcanization, were 20 sec., 30 sec. and 1, 2, 3 and 4 min. Details are given on the tables, together with physical test data. Under all conditions employed, satisfactorily cured mouldings were produced although it is possible, by examination of the tables, to note that there are in fact variations in the state of cure which follow the normal pattern of time and temperature dependence.

Compound 57082/1563–Breon nitrile 1001
(Tables 11.9, 11.10, 11.11, 11.12, 11.13).

Thirty-four injection mouldings were produced under conditions much as for compound 57082/1562 and examined fully for physical properties. Again no substantially undercured mouldings resulted and the physical test data bears out the regularity of the procedure.

Compound 57082/1565–Breon Polyblend 503
(Tables 11.14, 11.15).

This is a natural coloured compound with roughly equal proportions of

145

Fig. 11.5. The hopper, barrel and mould on the injection
moulding machine.

pvc and nitrile rubber. Rather more difficulty was experienced in
obtaining untorn mouldings on removal owing to the very weak tear
strength of the compound at the high moulding temperatures. It was
found better to achieve a rather higher state of cure so as to leave the
flowerpot in the female half of the tool on opening. This retention is
perhaps due to the higher strength of the sprue when relatively more fully
vulcanized. Twelve mouldings out of a total of 32 produced were
actually examined.

TRIAL COMPRESSION MOULDINGS

The three compounds were compression moulded for 5, 10, 15 and
20 min. at 153°C and 2, 4, 6 and 8 min. at 185°C.
(Tables 11.2, 11.3).

11.5 DISCUSSION OF RESULTS

The most interesting result was the general ease with which satisfactorily
vulcanized mouldings were produced almost, as it seemed, independently
of the setting of the machine. This 'easy' success is of course based on
previous experience in the development of the machine. For example,
the nozzle diameter has been arrived at as a good compromise between
the extremes of a stiff, scorchy, stock and a fluid slow curing stock; and
the experimental proof that there was no need to try a different sized
nozzle in handling stocks of such a wide spread in unvulcanized
properties, is a very good indication of the satisfactory design of the
machine. There appeared to be plenty of power available to inject the
stiffer stock.

146

11.5.1 BREON NITRILE 1001 COMPOUND

Looking at the results in more detail the trends in 300% modulus can be taken as a good indication of the state of cure of the compounds. 57082/1563 reaches a satisfactory 300% modulus in a 20 sec. cycle at 205°C, and in a 2 min. cycle at 188°C. Comparing with the 300% modulus achieved in compression moulding at 153°C, a time of 10 min. is required to reach the same modulus level.

11.5.2 BREON NITRILE 1042 COMPOUND

Very much the same results for injection mouldings are obtained with the other straight nitrile compound 57082/1562 for 2 min. at 188°C and 20 sec. at 205°C.

The 300% modulus results for vulcanizates prepared by compression moulding at 185°C indicate a reasonable state of cure in about 2 min.

11.5.3 BREON POLYBLEND 503 COMPOUND

The 300% modulus results for the compound 57082/1565 at temperatures in the mould of 177°, 188° and 199°C are very similar in pattern to the previously discussed 1042 compound, and a substantially full cure is reached in a 1 min. cycle at 199°C.

11.5.4 CURE RATE

From the results, for all three compounds, it can be seen that the conditions of vulcanization were such that a substantially full cure was obtained in all cases. Raising the temperature by 10°C is normally recognized as being equivalent to half the cure time. This pattern is in fact reproduced, 10 min. at 153°C giving equivalent results to 20-sec. at 205°C. This study was developed in later curometer tests and a regular vulcanization to temperature relationship was confirmed.

11.6 SUMMARY AND CONCLUSIONS

1 The successful injection moulding of high Mooney and low Mooney straight nitrile compounds and an ultra low Mooney high pvc ratio nitrile rubber blend is not difficult and has been carried out over a wide range of mould temperatures and times from 20 sec. at 205°C to 4 min. at 188°C.
2 The screw type machine has advantages in respect of good mechanical efficiency and general convenience.
3 Fully automatic feed and ejection can be readily integrated into the cycling controls so as to produce work on a push button basis with vulcanization cycle times of less than 30 sec.
4 The vulcanizate properties are generally equal to or slightly superior to compression mouldings.
5 The cost savings in injection moulding nitrile rubber compounds considerably narrow the differential in finished component cost as against cheaper polymers.

Note: The following additional paper by Mr Dale may be of interest: Short high-temperature cures with nitrile rubber compounds, *Rubber Journal* Dec., 1966, p.58.

Table 11.2 Compression moulding of compounds cure at 153°C

Compound 57082/1563 — Breon nitrile 1001

Cure time min.	Tensile strength lb/in²	Elongation at break %	100%	Modulus lb/in² 200%	300%	400%	Hardness °BS
5	2 660	370	585	1 490	2 275	–	67
10	2 610	325	575	1 605	2 420	–	69
15	2 610	305	630	1 700	2·540	–	69
20	2 740	310	700	1 740	2 610	–	70

Compound 57082/1562 — Breon nitrile 1042

5	2 290	405	438	1 195	1 850	2 210	61
10	2 390	435	405	990	1 795	2 260	63
15	2 320	415	400	995	1 690	2 205	63
20	2 430	450	400	1 015	1 840	2 210	64

Compound 57082/1565· — Breon Polyblend 503

5	1 870	320	485	1 050	1 760	–	67
10	2 470	360	575	1 100	1 600	–	67
15	2 060	320	595	1 030	1 810	–	68

Table 11.3 Compression moulding of compounds cure at 185°C

Compound 57082/1563 — Breon nitrile 1042

Cure time min.	Tensile strength lb/in²	Elongation at break %	100%	Modulus lb/in² 200%	300%	400%	Hardness °BS
2	2 710	310	617	1 790	2 660	–	69
4	2 740	330	778	1 815	2 640	–	72
6	2 840	310	685	1 985	2 800	–	73
8	2 630	300	660	1 655	2 560	–	72

Compound 57082/1562 — Breon nitrile 1001

2	2 450	470	280	845	1 760	2 100	60
4	2 490	430	439	1 067	1 590	2 405	60
6	2 460	430	364	992	1 755	2 280	62
8	2 500	460	379	1 000	1 685	2 220	63

Compound 57082/1565 — Breon Polyblend 503

2	1 750	280	790	1 220	–	–	67
4	1 600	300	700	1 090	1 600	–	67
6	2 030	325	720	1 180	1 760	–	69
8	1 800	310	670	1 160	1 760	–	70

Table 11.9 Injection moulding of compounds cure at 177°C

Compound 57082/1563   Breon nitrile 1001

Cycle time min.	Tensile strength lb/in^2	Elongation at break %	100%	Modulus lb/in^2		400%	Hardness °BS
				200%	300%		
1	2 762	357	495	1 430	2 500	–	68
2	2 655	305	662	1 827	2 600	–	68
3	2 600	300	658	1 635	2 600	–	71
4	2 785	305	604	1 604	2 700	–	71

Table·11.4 Injection moulding of compounds cure at 177°C

Compound 57082/1562   Breon nitrile 1042

Cycle time min.	Tensile strength lb/in^2	Elongation at break %	100%	Modulus lb/in^2		400%	Hardness °BS
				200%	300%		
1	2 603	457	273	1 001	1 736	2 407	62
3	2 606	416	288	1 151	1 977	2 448	62
4	2 740	406	300	1 009	1 865	2 570	63

Table 11.14 Injection moulding of compounds cure at 177°C

Compound 57082/1565   Breon Polyblend 503

Cycle time min.	Tensile strength lb/in^2	Elongation at break %	100%	Modulus lb/in^2		400%	Hardness °BS
				200%	300%		
1	2 108	355	729	1 155	1 714	–	67
2	1 928	330	620	1 081	1 718	–	68
3	2 184	330	822	1 255	1 890	–	69
4	2 202	325	776	1 225	1 900	–	68

Table 11.10 Injection moulding of compounds cure at 188°C

Compound 57082/1563   Breon nitrile 1001

Cycle time min.	Tensile strength lb/in^2	Elongation at break %	100%	Modulus lb/in^2		400%	Hardness °BS
				200%	300%		
30 sec.	2 874	477	349	1 005	1 795	2 648	63
1	3 033	393	359	1 352	2 255	–	65
2	2 904	315	648	1 813	2 756	–	68
3	2 785	300	519	1 676	2 785	–	69
4	3 144	336	603	1 753	2 951	–	70

Table 11.5 Injection moulding of compounds cure at 188°C

Compound 57082/1562   Breon nitrile 1042

Cycle time min.	Tensile strength lb/in²	Elongation at break %	100%	Modulus lb/in² 200%	300%	400%	Hardness °BS
1	2 692	442	284	845	1 623	2 250	62
2	2 708	427	257	884	1 608	2 438	61
3	2 747	446	260	890	1 812	2 510	62
4	2 776	440	266	1 901	1 901	2 566	61

Table 11.15 Injection moulding of compounds cure at 188°C

Compound 57082/1565   Breon Polyblend 503

Cycle time min.	Tensile strength lb/in²	Elongation at break %	100%	Modulus lb/in² 200%	300%	400%	Hardness °BS
1	2 197	370	711	1 098	1 677	–	66
2	2 174	340	725	1 173	1 795	–	66
3	2 176	340	747	1 186	1 829	–	67
4	2 188	350	813	1 242	1 823	–	67

Table 11.11 Injection moulding of compounds cure at 193°C

Compound 57082/1563   Breon nitrile 1001

Cycle time	Tensile strength lb/in²	Elongation at break %	100%	Modulus lb/in² 200%	300%	400%	Hardness °BS
10 sec. pressure	30 sec.	2 673	358	510	1 430	2 343		66

Table 11.6 Injection moulding of compounds cure at 193°C

Coumpound 57082/1562   Breon nitrile 1042

Cycle time	Tensile strength lb/in²	Elongation at break %	100%	Modulus lb/in² 200%	300%	400%	Hardness °BS
10 sec. pressure	30 sec	2 606	445	375	1 035	1 820	2 388	62

Table 11.12 *Injection moulding of compounds cure at 199°C*   Breon nitrile 1001

Compound 57082/1563

Cycle time	Tensile strength lb/in²	Elongation at break %	100%	Modulus lb/in² 200%	300%	400%	Hardness °BS
20 sec. pressure	40 sec.	2 572	300	652	1 748	2 572	–	68
20 sec. pressure	45 sec.	3 000	354	593	1 649	2 639	–	68

Table 11.7 *Injection moulding of compounds cure at 199°C*

Compound 57082/1562   Breon nitrile 1042

Cycle time	Tensile strength lb/in²	Elongation at break %	100%	Modulus lb/in² 200%	300%	400%	Hardness °BS
20 sec. pressure	30 sec.	2 797	450	381	1 100	1 958	2 555	61

Table 11.16 *Injection moulding of compounds cure at 199°C*

Compound 57082/1565   Breon Polyblend 503

Cycle time min.	Tensile strength lb/in²	Elongation at break %	100%	Modulus lb/in² 200%	300%	400%	Hardness °BS
1	2 282	365	818	1 212	1 760	–	66
2	2 310	360	813	1 190	1 771	–	68
3	2 233	350	796	1 230	1 805	–	72
4	2 243	335	850	1 325	1 926	–	71

Table 11.13 *Injection moulding of compounds cure at 205°C*

Compound 57082/1563   Breon nitrile 1001

Cycle time	Tensile strength lb/in²	Elongation at break %	100%	Modulus lb/in² 200%	300%	400%	Hardness °BS
20-sec. pressure	20 sec.	2 814	326	680	1 504	2 650	–	68

Table 11.8 Injection moulding of compounds cure at 205°C

Compound 57082/1562   Breon nitrile 1042

	Cycle time	Tensile strength lb/in^2	Elongation at break %	100%	Modulus lb/in^2		400%	Hardness °BS
					200%	300%		
20-sec. pressure	20 sec.	2 650	484	303	940	1 780	2 340	61

DISCUSSION

M. E. GRINDLE *(Dowty Seals)*: A particular problem in high temperature moulding, e.g. injection moulding, is increased sulphur spotting in finished components. Could Mr Dale comment on ways of reducing or overcoming this?

ANSWER: Sulphur spotting in finished components can be minimized by ensuring complete dispersion of the sulphur in the mix. It is always best with nitrile compounds to put the sulphur in very early in the mixing cycle. We have found that a treated sulphur such as Anchor Chemicals M.C. sulphur does assist to achieve the best dispersion. Also one can try a low sulphur curing system or one which is completely sulphur free.

L. M. GLANVILLE *(International Sythetic Rubber Co. Ltd)*: It is assumed that many products based on nitrile polymers require good compression set characteristics. Has Mr Dale any comparative data comparing injection and compression methods for this property?

A: We have, in fact, checked compression set behaviour of injection moulded nitrile rubber compounds against the same compounds when compression moulded. Results are given in an article of mine, *Rubber Journal* December 1966, and there was a clear indication that the injection mouldings were much more susceptible to set.

Of course, in line with all other rubber technology it is possible to adjust techniques and formulations to overcome this behaviour and, for example, a short oven post cure is a practical technique for improving compression set.

R. F. POWELL *(Trist Moulding and Seals)*: Should screw speeds be as low as possible to reduce static stock time in the barrel for any given cycle time?

A: Since nitrile compounds are generally insensitive to the times and temperatures generally resulting when they pass through the screw and barrel of an injection moulding machine, there is normally no need to consider the effect of screw speed on the dwell time in the barrel. Going to the extreme, it would be possible to scorch the compound by frictional heat if the screw speed was too high, so one would always advise a reasonably low screw speed.

CHAPTER 12

Fundamental aspects of injection moulding butyl rubber

D. A. BOOTH, A.N.C.R.T., A.I.R.I.
Esso Research SA

12.1 INTRODUCTION

THE injection moulding of rubber is not a new process. References to this invention may be found as early as 1940 and experimental work must have been started before this date. However, over the past six years there has been a considerable surge of interest in the process and the number of injection moulding machines manufactured specifically for rubber has increased remarkably (Fig. 12.1).

Surveys carried out in the past have shown that in October 1963, there were 47 rubber injection machines available. Of this total only six were simple ram types, the majority being of reciprocating screw construction. In the 14 months following October 1963 the number of rubber machines on the European market doubled and by the beginning of 1965 there were about 100 models being offered for sale (only nine or ten of these were ram types). From 1965 to the present day (1968) the growth of interest in rubber injection moulding has continued and has manifested itself in two ways. First, the machinery manufacturers have continued to adapt and improve machine designs for elastomers and have placed more machine models on the world market. Secondly, the number of rubber injection demonstrations at the major polymer and engineering exhibitions has increased, as has the number of articles written on the subject in the trade press. Currently there are some 170 machine models available on the market for rubber.

As a result of this activity, the number of machines in commercial use has risen sharply. It is always difficult to put an accurate figure to a commercial usage of this type, but it has been estimated that there are about 300 machines now producing rubber in the United Kingdom.

The work described in this chapter has been performed using reciprocating screw machines both in the Esso Research Laboratories and at different machine manufacturers in Europe.

12.2 TEMPERATURE VARIABLES

The mode of operation of the screw machine (Fig. 12.2) is that rubber (in strip or granular form) is fed to the rear end of the screw. The screw passes stock forward towards the nozzle end of the injection barrel, and the screw is forced to the rear. When a predetermined amount of stock has been prepared in this fashion the screw ceases to rotate. Injection is accomplished by the screw moving bodily forwards under hydraulic pressure—the rubber being forced through the nozzle and

The photographs illustrating this chapter, Figs 12.20, 21, 23, and 32 will be found on Plates 15 and 16 (between pages 134 and 135).

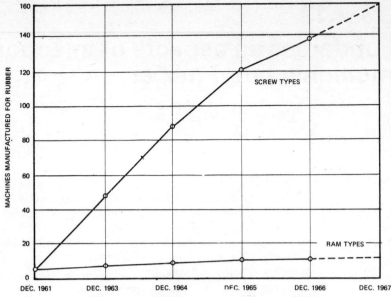

Fig. 12.1. World increase in rubber machines.

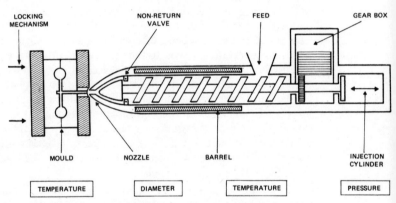

Fig. 12.2. Reciprocating screw rubber injection moulding machine.

into the mould.

It is recognized that injection moulding has advantages particularly that of giving a short cure. Fig. 12.3 indicates how these advantages come about. The upper diagram shows compression moulding—temperature against curing time. There are four cure state areas, labelled A to D (start of scorch to reversion). It can be seen that, in order adequately to cure the interior of a moulding, the surface has entered the reversion area. Curing time is of the order of 30 min.

The lower diagram shows injection moulding. The stock temperature is raised and controlled in stages: (1) is preheating of stock by a preplasticizing unit; (2) is accumulation of the shot at the front end of the screw, and (3) is heat rise on injection due to the nozzle restriction. It can be seen that the material enters the mould at a quite high temperature, and that the difference between the state

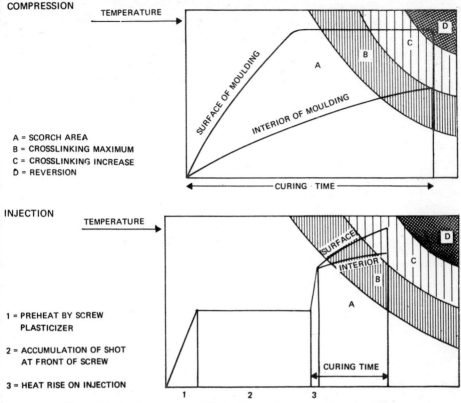

COMPRESSION

TEMPERATURE

A = SCORCH AREA
B = CROSSLINKING MAXIMUM
C = CROSSLINKING INCREASE
D = REVERSION

SURFACE OF MOULDING

INTERIOR OF MOULDING

CURING TIME

INJECTION

TEMPERATURE

1 = PREHEAT BY SCREW
    PLASTICIZER

2 = ACCUMULATION OF SHOT
    AT FRONT OF SCREW

3 = HEAT RISE ON INJECTION

SURFACE

INTERIOR

CURING TIME

Fig. 12.3. Curing aspects of injection and compression moulding.

of cure of the interior, and exterior of the moulding is thus minimized.
Owing to the fact that stages (1) and (2) take place during the
previous curing cycle, the curing time is very short (45 sec.).

Typically for butyl, one can turn the injection cylinder at 110°C. The
material will reach the front end of the screw at about 120°C.
Injection will give a heat rise of some 40–50°C, which will lead to
material entering the mould at around 150–160°C. This is already
a curing temperature by conventional standards.

The ideal rubber compound for injection moulding should have good
scorch safety so that high preinjection barrel temperatures may be
used to advantage; it should flow very easily through narrow apertures
and runner systems, giving rapid mould filling with high heat build-up
during injection; on reaching the mould it should cure quickly, giving
a stable, reversion resistant vulcanizate at the high mould temperatures
used.

12.3 BASIC FLOW CHARACTERISTICS

12.3.1 POLYMER EFFECTS

For comparative purposes basic flow characteristics were determined
for various rubbers compounded with 40 parts of various fillers.
Master batches, consisting of elastomer and filler only, were injected

155

freely through a standard nozzle into an insulated container. Injection times were measured in sec. and the temperature of the stock after injection was obtained by a thermocouple inserted into the polymer mass.

Similar trends were observed for all the fillers examined and thus the trends will be illustrated using results obtained for SRF black. The injection pressure/flow time relationships for the polymers are shown in Fig. 12.4. Injection gauge pressure is shown on the vertical axis. Actual pressure on the material at the nozzle can be found by

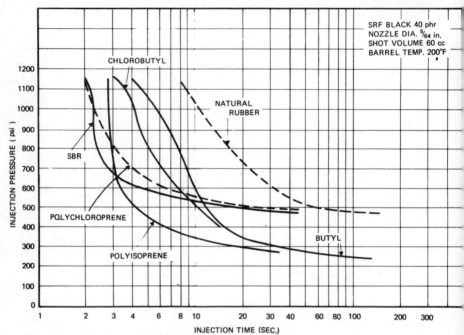

Fig. 12.4. Injection moulding rubber. Flow characteristics—comparison of polymers injection moulding vs flow time.

multiplying by 13·6 (the area ratio of the injection cylinder and the screw being 13·6 to 1). Clearly, SBR and polychloroprene flow very readily under high injection pressure conditions, but become progressively harder to inject as pressure decreases. Polyisoprene also flows readily and maintains its advantage over the whole pressure range. The steeper curve shown by polyisoprene and butyl is to be preferred, since the continued ease of flow under low pressure conditions is important when complicated multi-cavity tools with involved runner systems are to be used, because the pressure drop in the feed system due to the length of the runners does not then lead to a significant reduction in injection speed.

Natural rubber is shown to be considerably slower than the other polymers over the whole pressure range. The Mooney viscosities of all the batches were comparable.

The heat build-up on injection curves (Fig. 12.5) show less difference between the polymers than might have been expected considering the differences observed in flow characteristics. Butyl and chlorobutyl show an advantage at the higher end of the pressure range and natural

156

rubber gives the lowest heat build-up. Generally, over most of the injection pressure range the heat rise on injection for all rubbers falls within a 11°C spread.

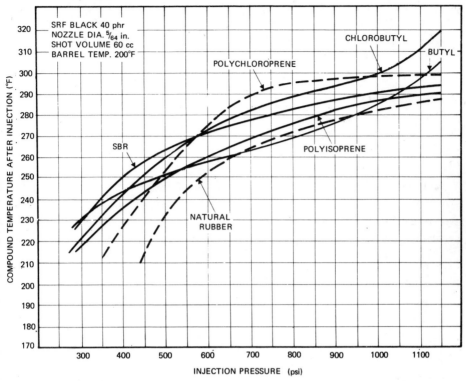

Fig. 12.5. Injection moulding rubber. Temperature rise on injection.

12.3.2 TEMPERATURE EFFECTS

Flow chacterizations were repeated at various barrel temperatures so that the temperature dependency of flow behaviour could be determined (Fig. 12.6).

Natural rubber is the most temperature dependent polymer and SBR, polychloroprene, and the lower viscosity butyl type rubbers are least affected. These last three were indistinguishable from one another and are represented by a single line for clarity. The Mooney viscosities of the master batches are given in Fig. 12.6.

12.3.3 VISCOSITY EFFECTS

A change in the viscosity of the material being injected must alter both the flow rate and the shear forces experience at the nozzle (and hence the heat build-up on injection).

Two polymer types are shown, butyl rubber (variation of Mooney viscosity with grade of polymer) and natural rubber (variation of Mooney viscosity by mechanical breakdown). Natural rubber is also compared with a poly*iso*prene master batch of the same viscosity. It can be seen that, although natural rubber and poly*iso*prene exhibit the same shape of curve, the synthetic rubber flows much more easily (Fig. 12.7). The butyl master batch (Mooney 76) compares favourably with the

157

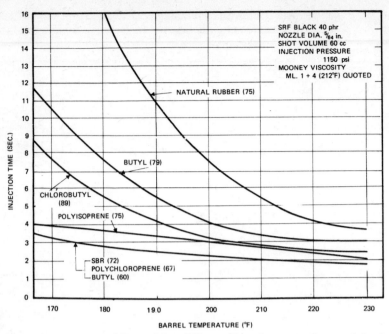

Fig. 12.6. Injection moulding rubber. Effect of barrel temperature on injection time.

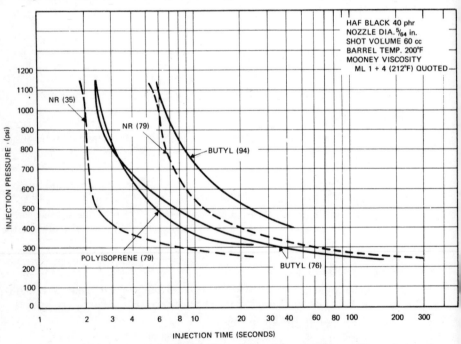

Fig. 12.7. Injection moulding rubber. Flow characteristics–effect of viscosity.

poly*iso*prene (Mooney 79), and the butyl (Mooney 94) compares
reasonably with the natural rubber (Mooney 79).

The low viscosity natural rubber (Mooney 35) flows very readily, but
obviously the shear forces at the nozzle have been much reduced, since

158

the heat build-up on injection is appreciably lower than the higher viscosity master batch. (Fig. 12.8). Conversely, the lower viscosity butyl master batch both flows more easily than that of the higher viscosity one and *also* gives an increase in stock temperature after injection.

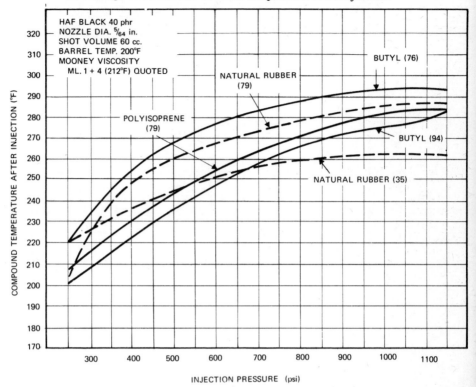

Fig. 12.8. Injection moulding rubber. Temperature rise on injection–effect of viscosity.

12.3.4 PLASTICIZER VARIATIONS

Further investigations into the effects of viscosity on injection characteristics were made by taking three different types of plasticizer and adding them in varying proportions to butyl rubber base compounds. Compound viscosity, injection speed, and temperature rise on injection were determined (using constant machine conditions) for the base compounds containing 0–20 phr of a paraffinic mineral oil (Esso Process Oil V), a paraffin wax, and an ester plasticizer (di-tridecyl phthalate). Figs 12.9, 12.10 and 12.11 show the decrease in Mooney viscosity experienced with increasing amounts of each plasticizer in base compounds consisting of 100 parts of butyl rubber and 60 parts of carbon black. The base formulations are indicated in the figures.

A shot volume to 60 cc. was used throughout the experiments together with a primary injection pressure of 1 150 psi on the reciprocating screw (15 640 psi on the material at the nozzle). A barrel temperature of 93°C, and a standard nozzle of final aperture ⁵⁄₆₄ in. diam. Fig. 12.12 shows the decrease in injection times experienced with decreasing viscosity.

The injected masses of rubber compound were caught in an insulated container, and the mass temperatures were determined by means of a

159

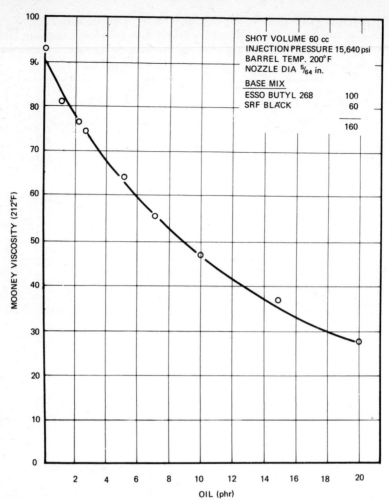

Fig. 12.9. Mooney viscosity vs oil loading.

suitable thermocouple probe. It can be seen from Fig. 12.13 that definite peaks exist in the mass temperature/viscosity curves produced by the three plasticizers. The temperature rise at the peak for the oil filled mix was 42°C.

12.4 NOZZLE SIZE VARIATIONS

The heat build-up at the nozzle is a function of the shearing forces experienced by the rubber at the nozzle. The nature of the compound ingredients, compound viscosity, injection speed, barrel temperature, and nozzle aperture size all contribute to this effect. It has been demonstrated that a maximum can be obtained in the heat build up on injection curve. Maximum heat build-up at the nozzle is desirable (provided scorching does not take place) because this leads to faster curing cycles. It is obvious that, if the scorch characteristics of the polymer/formulation permit, a high cylinder temperature 'platform' will provide a basis from which a maximum heat rise on injection (through a suitable nozzle) will lead to the melt entering the mould at a curing temperature. It can thus be seen that polymer type (scorch characteristics) and nozzle design are important parameters for consideration.

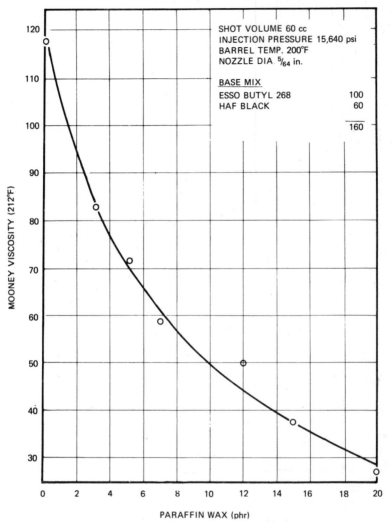

Fig. 12.10. Mooney viscosity vs *wax loading.*

Nozzle size then, is of great interest to injection moulders. It is obvious that a smaller than standard nozzle will cause a decrease in injection speed. Increasing the barrel temperature counteracts this tendency to some extent, but within the range of temperatures examined in this series of experiments there was no overlap (Fig. 12.14).

However, when the results of the heat rise on injection against inection pressure experiments were examined (Fig. 12.15), it could be seen that the smaller nozzle might offer advantages in terms of increased melt temperature and faster curing. This whole concept was then investigated further during a commercial moulding run, using a 40 Shore butyl compound and 5 nozzles having different final diameters (Table 12.1). Injection characteristics are shown in Fig. 12.16.

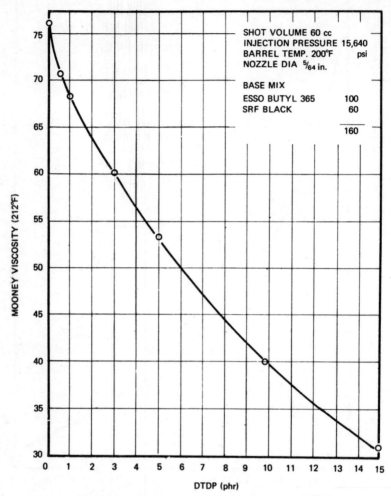

Fig. 12.11. *Mooney viscosity* vs *phthalate loading.*

162

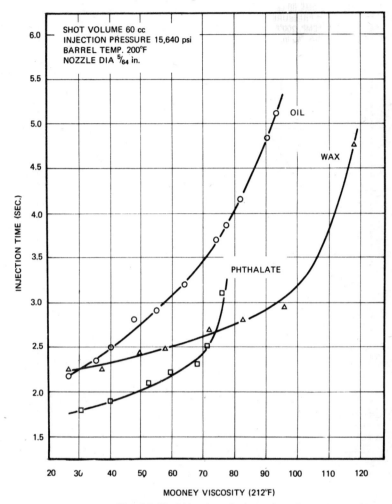

Fig. 12.12. *Injection time* vs *viscosity.*

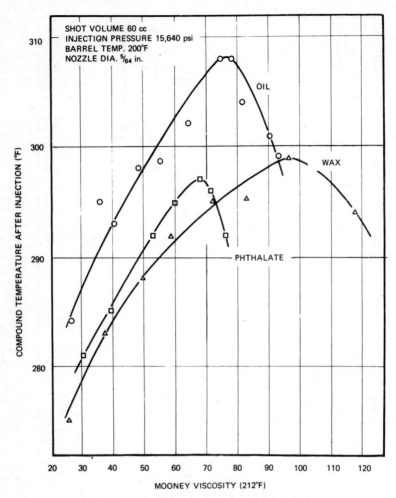

Fig. 12.13. *Temperature rise on injection.*

164

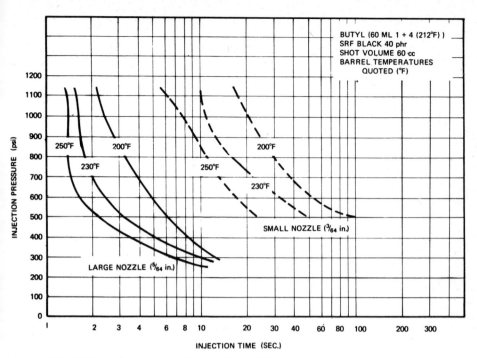

Fig. 12.14. Injection moulding rubber. Effect of nozzle size on injection time.

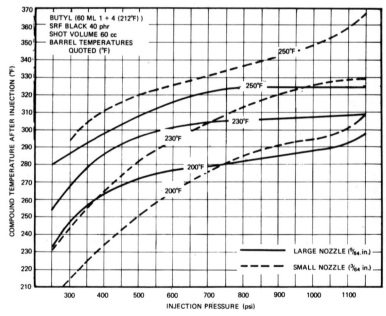

Fig. 12.15 Injection moulding rubber Temperature rise on injection—effect of nozzle size.

Table 12.1 40 Shore A butyl compound

Esso Butyl 218	100
HAF	35
FEF	5
M. 100 clay	15
Oil	25
ZnO	5
Sulphur	1·5
MBTS	0·5
ZDC	2
	189·0

Physical properties (30min. at 320°F)

Hardness (°BS)	40
Tensile strength (psi)	2 090
300% Modulus (psi)	610
Elongation (%)	600
Compression Set B (22 hr. at 70°C)	22·9

Nozzle diameters

$\frac{3}{64}$ in., $\frac{1}{16}$ in., $\frac{5}{64}$ in., $\frac{3}{32}$ in. and $\frac{1}{8}$ in.

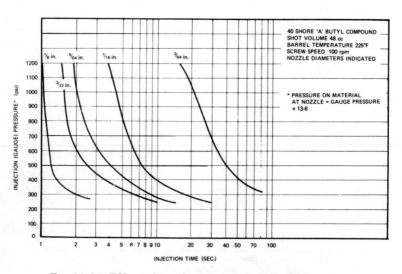

Fig. 12.16. Effect of nozzle size on injection characteristics.

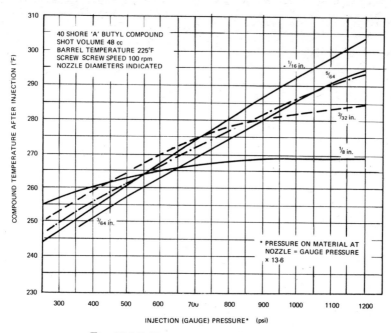

Fig. 12.17. Temperature rise on injection.

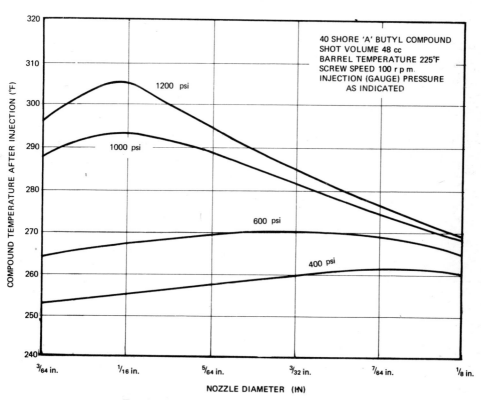

Fig. 12.18. Temperature rise/pressure sensitivity.

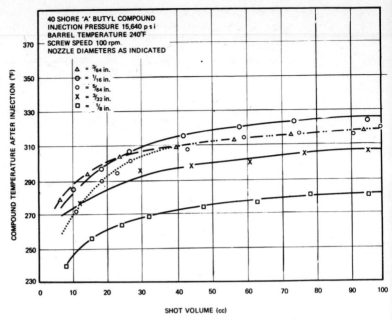

Fig. 12.19. Shot volume vs *temperature rise.*

The temperatures of the compound after injection were also studied (Fig. 12.17). The injection time/pressure relationships were found to be consistent with previously reported work for butyl rubber, and the temperature/pressure relationships indicate that considerable scope exists in the matter of correct choice of nozzle for compound viscosity characteristics and injection pressure. Replotting the temperature results against nozzle diameter (at constant pressure) highlights the fact that, for a given compound, a maximum heat rise on injection can be isolated, and that changes in pressure appear to shift this maximum position with respect to nozzle size (Fig. 12.18).

The effect of shot volume on compound temperature after injection was also determined for the same range of nozzles as used previously. Fig. 12.19, shows that temperature tends to increase with shot volume. This effect is found to be less significant if over 50% of the available volume (a practical economic point) is considered as being the 'normal' volume range for the machine.

12.5 SIZES OF GATES AND RUNNERS

It has already been stated that heat rise at the nozzle must depend on speed of injection and on compound viscosity. Heat must also be gained by the compound as it travels along the runner system and through the gates into the cavities.

In order to determine the effects of various sizes and configurations of runners and gates (as applied to a standard cavity), a mould consisting of three rotatable inserts was made. This mould is shown in Fig. 12.20, and a moulding produced from it in Fig. 12.21 (see Plate 15). Using this moulding, it is possible to present two different sized runners (by rotation of the middle insert) to two standard cavities. By rotation of the cavities, it is possible to change the type of gate or entry point in a similar manner.

Many experimental runs were made, and the conclusions drawn from these can be illustrated in Fig. 12.22 and Table 12.2 which show results

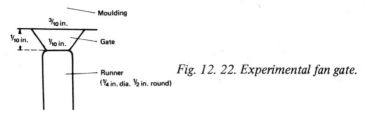

Fig. 12. 22. Experimental fan gate.

Table 12.2 Effect of fan gate depth

Runner size (in.)	Fan gate depth (in.)	Injection time (sec.)	Vulcanizing time (sec.)
¼ (½ round)	0·010	5·4	35
¼ (½ round)	0·020	2·8	35
¼ (½ round)	0·030	2·1	35
¼ (½ round)	0·040	2·0	30

40 Shore A butyl compound
Shot volume 48 cc.
Mould temperature $400°F$
Barrel temperature $235°F$
Screw speed 100 rpm
Injection pressure on material 15 640 psi
Nozzle diameter ⁵⁄₆₄ in.

using a ¼ in. (½ in. round) runner combined with four different depth fan gates (10, 20, 30 and 40 thou. deep). For the same 40 Shore compound as used previously, it can be shown that (under standard conditions) injection time increases as gate depth decreases, and that vulcanizing time can be measurably affected by selection of gate depth.

Vulcanizing time was defined as the minimum cycle time necessary for porosity to disappear within the moulding (a disc of 3 in. diameter, ³⁄₁₆ in. thick).

The experiment was repeated using different sized nozzles, and keeping the mould configuration and injection conditions constant.

For each nozzle, the free injection time was taken (injection freely to atmosphere, without the mould in place), The compound temperature after free injection was noted (by means of a thermocouple probe), and the vulcanizing time and injection time with the mould in position were taken (Table 12.3).

It can be seen that nozzle size affects free injection time and temperature rise on injection markedly. The moulding injection time is similarly dependent on nozzle diameter, and, even though a lower than usual barrel temperature is used ($107°C$), scorching occurs in the

169

Table 12.3 Effect of nozzle size

Nozzle size (diam.) (in.)	Free injection time (sec.)	Compound temperature after injection (°F)	Vulcanizing time (sec.)	Mould injection time (sec.)
3/64	16·3	295	< 35*	23·5*
1/16	3·8	304	35	5·3
5/64	1·9	294	35	2·9
3/32	1·5	285	35	2·2
1/8	1·0	270	45	1·3

40 Shore A butyl compound
Shot volume 48 cc.
Barrel temperature 225°F
Screw speed 100 rpm
Injection pressure on material 16 320 psi
Mould temperature 400°F
Fan gate 0·040 in. deep ¼ in. (½ round) runner

* Scorch took place before mould filling was complete.

cavity (owing to slow injection) with the smallest nozzle in place. Injection becomes easier as nozzle diameter increases, but compound temperature falls also. As a result, no differences can be detected in vulcanizing time until the largest nozzle is used. This nozzle produces fast injection with a significantly lower temperature than the others, and because of this the curing time was extended.

12.6 FEEDSTOCK FORMS

Rubber injection moulding is becoming more accepted in both the rubber and the plastics industries. These two industries tend to differ in their basic feedstock forms (i.e. strip versus granules). With this in mind, determinations were made of the effects of various feedstock forms on plasticizing rates at different screw speeds and barrel temperatures, using a typical medium hardness butyl compound (Table 12.4). A standard nozzle (5/64 in. diam.) was used throughout.

The compound was finalized and sheeted off ¼ in. thick. Strip was cut from the sheet (1½ in. wide) and various sizes of granules were also prepared by cutting from the same sheet. Fig. 12.23 (Plate 16) shows the feedstock produced: (1) is the granular product of the Masson cutter (small irregular granules), (2) is the ¼ in. cube, (3) is the ½ in. × ½ in. × ¼ in. granule, and (4) is the 1 in. × 1 in. × ¼ in. size. The granules were cut, dusted with talc, and stored in 4 lb. tins. No trouble was experienced with their flowing together over a period of 2–3 months at room temperature and they proved to be readily usuable and free flowing after this time.

The time was recorded for the screw to retract during its preparation of a 52 cc. shot volume. This particular volume was chosen simply because it suited the mould which was in position on the machine at the time. From this figure the number of grammes of material being plasticized per sec. was calculated and hence the plasticizing rate in kgm./hr. (assuming continuous plasticization).

The feedstocks were fed to the injection moulding machine using a screw speed of 100 rpm and a barrel temperature of 93°C (Fig. 12.24). It can be seen that feedstock form does affect plasticizing rate. Strip

Table 12.4 Medium hardness butyl compound

Esso Butyl 268	100
HAF	30
FEF	10
M. 100 clay	10
Petrolatum	10
P. Wax	20
Zinc stearate	2
Zinc oxide	5
Sulphur	1·5
MBTS	0·5
ZDC	2
	——
	191
	——

Physical properties (30 min. at 320°F)

Hardness (Shore·A°)	56
Tensile strength (psi)	1 920
300% Modulus (psi)	610
Elongation (%)	650
Compression set (%)	
ASTM B (22 hr. at 70°C)	23·5
Mooney scorch (MS 250°F)	16·5
Mooney viscosity	
(ML 1 + 4 212°F)	31

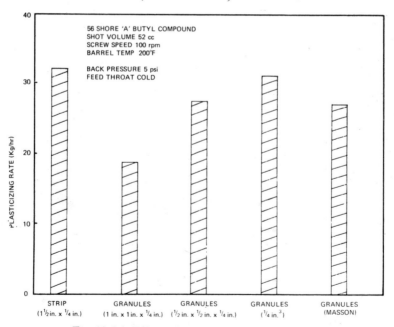

Fig. 12.24. Effect of feed type on plasticizing rate.

171

feed is the most efficient under these conditions with the ¼ in. cubical granules running a close second. Heat build-up on injection is not significantly affected by feedstock type.

The effect of screw speed was determined on plasticizing rate for strip, ¼ in. cubical granules, and ½ in. × ½ in. × ¼ in. granules (Fig. 12.25). It can be seen that below a speed of 110 rpm there

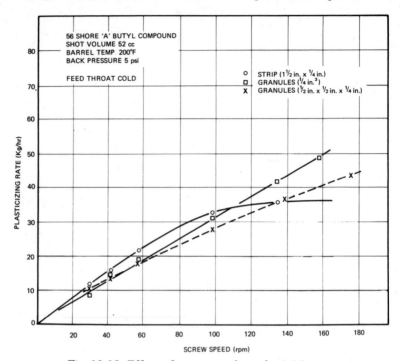

Fig. 12.25. Effect of screw speed on plasticizing rate.

exists a marginal advantage for the strip feed. However, above this speed, it can be clearly seen that the granules (particularly the ¼ in. cubes) feed considerably more efficiently than the strip rubber.

The determinations were repeated at different barrel temperatures (Fig. 12.26). Temperature did not significantly alter feeding rate. This is illustrated by reference to a granular feedstock.

12.7 NON-RETURN VALVES

Injection efficiency, and whether or not a screw tip non-return valve should be used for rubber injection moulding, have long been subjects of lively discussion among machine manufacturers. In order to test the applicability of such a non-return valve to butyl rubber formulations, a series of tests was made on an Ankerwerk V36–150 rubber machine.

Compounds covering a wide range of Mooney viscosity were prepared· (38 → 160 Mooney ML 1 + 4 100°C) and injected freely into atmosphere using both a plain spiral screw tip and a version including a non-return value (Fig. 12.27).

172

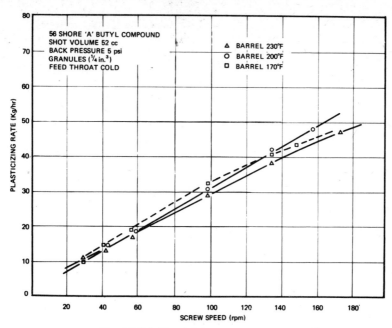

Fig. 12.26. Plasticizing rate vs screw speed.

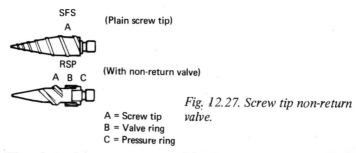

SFS
A
(Plain screw tip)

RSP
A B C
(With non-return valve)

Fig. 12.27. Screw tip non-return valve.

A = Screw tip
B = Valve ring
C = Pressure ring

When shot volumes were compared for the two screw tips, with machine conditions identical for both, it was seen (Fig. 12.28) that the non-return valve was increasing injection efficiency. The efficiency increase is greatest for the soft formulations and least for the hard compounds.

The major advantages of using a screw tip non-return valve have been found to be the resultant gain in shot volume control sensitivity and the decrease in sensitivity to compound viscosity change.

12.8 CURE SYSTEMS AND REVERSION

It is well known that for most elastomers it is possible to reach that state of vulcanization known as 'overcure'. At high temperatures particularly, a state of overcure cometimes becomes (on the surface) catastrophic. This state is known as reversion or degradation. It can be postulated that, during vulcanization, two reactions proceed concurrently; one a crosslinking reaction, and the other a degradative reaction. Normally crosslinking proceeds at a faster rate than degradation, but at very high temperatures the balance may be tipped in the opposite direction. As an

173

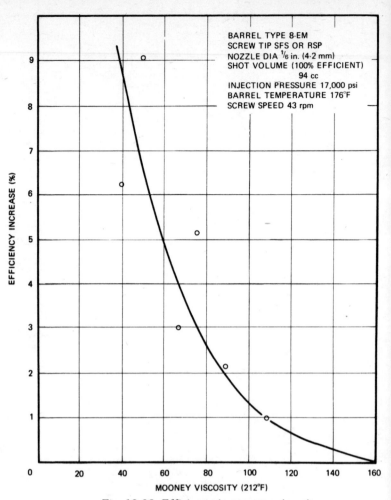

Fig. 12.28. *Efficiency increase* vs *viscosity*.

extreme case pyrolysis may be quoted as an example.

An investigation into the influence of curing temperatures and vulcanizing systems was carried out using simple butyl and chlorobutyl compounds.

The rubbers were injection moulded at three mould temperatures (182°C, 193°C, 210°C) using cure times ranging between 30 sec. and 4 min. The test moulding was a single cavity hollow truncated cone of wall thickness 1/10 in. and base thickness 1/4 in.

Tensile test pieces could be cut from the walls of the cone in order to compare the physical properties of injection moulded and conventionally press cured rubber compounds.

Variations of tensile strength and 300% modulus with cure time were plotted graphically. The compound recipe, the Mooney scorch, and the press cured physicals are all indicated in Fig. 12.29.

The butyl/sulphur systems is fast curing under injection moulding conditions giving good technical vulcanizates in under 45 sec.

However, both tensile strength and 300% modulus decrease with cure time at the highest mould temperature, and some reversion was detected visually after three min. at 210°C. The system is suitable for sections of about ¼ in. thickness and less, but where thicker sections are to be moulded rapidly, a more reversion-resistant system would be needed. This may be illustrated by contrasting the butyl-sulphur system with the same compound employing a resin cure (Fig. 12.30).

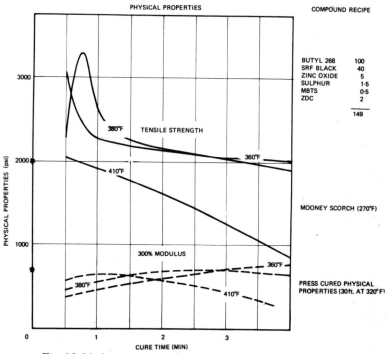

Fig. 12.29. Injection moulding butyl rubber.

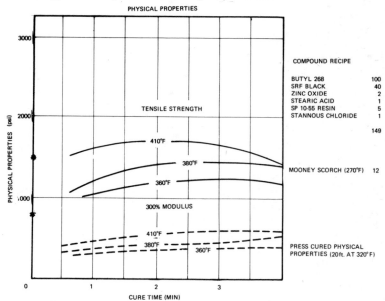

Fig. 12.30. Injection moulding butyl rubber.

175

The resin cured compound, although slower curing than the sulphur based system, gives very stable vulcanizates which improve in tensile strength and 300% modulus as both cure time and temperature are increased. One might say the higher the better.

The chlorobutyl system approaches the ideal curing characteristics mentioned previously (Fig. 12.31).

Both tensile strength and 300% modulus are fully developed after 30 sec. at all three mould temperatures and they are not affected by overcuring. No reversion occurs even after extended periods at high temperatures. This is attributed to the extreme stability of the oxide type cure.

It can be seen that the nature of the compound dictates the nature of the curing conditions.

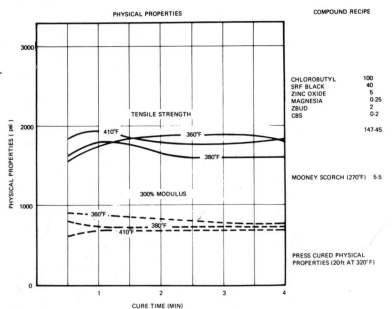

Fig. 12.31. Injection moulding chlorobutyl rubber.

12.9 COMPARISON BETWEEN COMPRESSION AND INJECTION MOULDING

In order to determine how the physical properties of a typical sulphur cured butyl compound varied under compression or injection moulding, a range of cures was performed at 160°C (compression) and at 204°C (injection).

The properties of standard compression moulded tensile pads were compared with the properties of samples cut from injection moulded 'flowerpots' and the properties of various injection and compression moulded test buttons were also examined. A test injection mould for various test buttons, such as hardness, Goodrich flexometer, Yersley oscillograph test pieces, and compression set buttons, was manufactured for this work (Fig. 12.32, Plate 16).

An examination of physical properties shows (Fig. 12.33) injection moulded modulus to be lower than the compression moulded modulus

176

by about 150-200 psi. Tensile strengths are higher for the injection moulded samples after fairly short cure times, but there is a sensitivity to prolonged curing. 'State of cure' tests (i.e. volume swell in

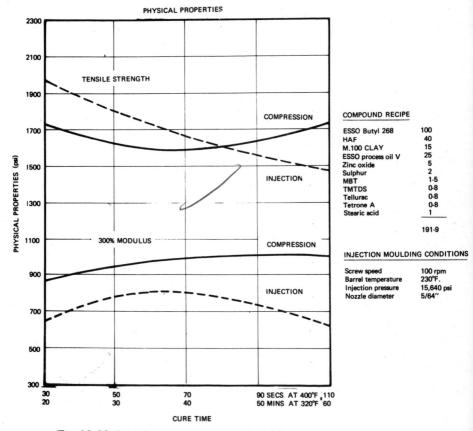

Fig. 12.33. Injection vs compression moulding.

cyclohexane — 72 hr. at 20°C) were carried out on tensile test pieces, taken from the tensile test runs (Fig. 12.34).

The results confirm that, as might be deduced from the 300% modulus values already quoted, the state of cure of the injection moulded samples is not as good as that of the compression moulded samples. The differences shown might be expected to be less, if a lower injection mould temperature than the one used (204°C) were employed. More reversion resistant cure systems would also have this effect.

Results on compression and injection moulded compression set buttons show the effect of state of cure quite clearly (Fig. 12.35). Set values by injection (204°C) are uniformly about 5% above the values obtained by compression (160°C). Production mouldings (being injected and placed in a tote bin) might be expected to show less variation from the compression moulded norm due to the mass of mouldings remaining at a higher than usual temperature over a longer period of time.

177

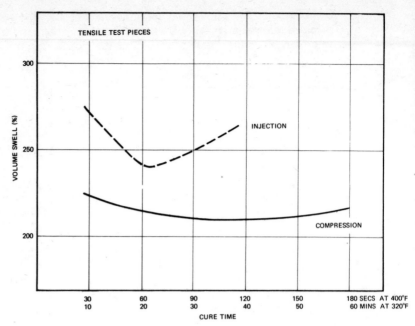

Fig. 12.34. Volume swell in cyclohexane.

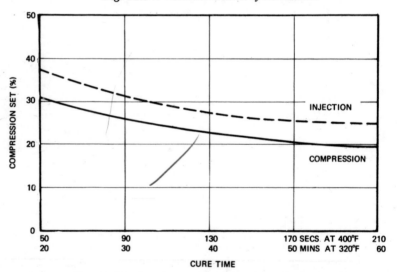

Fig. 12.35. Compression set (injection vs compression).

12.10 CONCLUSIONS

In conclusion, it can be said that all the common elastomers can be injection moulded, and that interest in the process continues to grow.

Viscosity and scorch characteristics, together with reversion resistance are the key factors involved.

Generally, small differences were observed in the physical properties of compression and injection moulded samples, there being a tendency for the injection moulded samples to have higher tensile strengths and lower moduli.

At high mould temperatures, it can be shown that the state of cure of injection moulded samples is not quite as high as that of conventionally compression moulded samples.

Butyl and chlorobutyl's extreme scorch safety, and easy injection characteristics provide a basis for building an almost ideal range of injection moulding formulations. Fast injection at high shear rates through carefully selected nozzle and gate sizes has been shown to lead to very high rubber melt temperatures. This, coupled with excellent heat stability, allows high mould temperatures to be used, and enables fast cures to be achieved.

Rapid plasticization of both strip rubber and granules at high screw speeds leads to quick preparation of heated stock prior to injection. Fast overall moulding cycles are a direct result of these facts.

The butyl type polymers are extremely easily injected, and they maintain their ease of flow over a wide temperature and pressure range.

Of the polymers evaluated, chlorobutyl demonstrated the ideal curing characteristics for injection moulding. Cure was fast and very resistant to reversion, and physical properties did not change with increases in either cure time or temperature.

DISCUSSION

P. H. KELLETT *(Shoe and Allied Trades Research Association)*: Has any work been done on rise of temperature obtained on passage of rubber through the gate into the mould, e.g. by using a simple extension on the nozzle to simulate runner and gate?

ANSWER: No, since it is extremely difficult to obtain and satisfactorily place a sufficiently sensitive temperature detector in the gate area. Problems would arise owing to (a) interference from the hot (200°C) mould, and (b) the transient nature of the phenomenon to be measured.

No extension nozzle (two constrictions), such as you describe, has been tried.

H. M. GARDNER *(T. H. and J. Daniels Ltd)*: Variations in nozzle diameter affect individual characteristics (temperature rise, injection time, etc.) —these are to some extent self-compensating. Does this eliminate the need for a range of nozzle diameters?

The short answer is no. You will remember that it was said 'to *some* extent self-compensating' with reference to injection speed *versus* heat build up with the mould in place.

It has been stressed that there is an optimum nozzle size for each compound under each set of machine conditions. A wide range of nozzle sizes should be available.

N. H. KING *(Firestone Tyre and Rubber Co.)*: Has any work been done on butyl compounds using sulphur curing systems to develop a compound resistant to reversion?

A: Work is continuing on sulphur cures for butyl and chlorobutyl, but I have nothing further to add to what has already been said at this stage.

M. D. HEAVEN *(Raychem Ltd)*: Would it not be possible to reduce injection times (and at the same time increase the amount of heat produced in the nozzle) by using 'implosion' moulding methods as sometimes used in conventional injection moulding?

A: 'Implosion' moulding is a possibility that has not yet been investigated to my knowledge. One might find one had scorch problems, however, if one tried to apply high pressures to the rubber melt without actually injecting it.

CHAPTER **13**
Injection moulding of Du Pont elastomers

J. C. BAMENT, A.I.R.I.
Du Pont Co. (U.K.) Ltd

13.1 INTRODUCTION

THE rapidly increasing number of injection moulding machines now operating in Europe and North America demonstrates that injection moulding is well established and becoming of ever increasing importance in the industry. The advent of this technique as a regular production process demands more precise control of elastomer compounding and processing conditions than for compression moulding. This chapter therefore presents information on machines of both types with which we have had experience, and discusses the special compounding measures necessary in the injection moulding of Neoprene, Hypalon synthetic rubber, and Viton fluoroelastomer. Also included is a list of possible remedies for everyday production problems.

The technical and economic advantages of injection moulding rubber mechanicals have been widely discussed, during the last two years [1,2,3]. However, volume production of a part appears essential to amortize the increased equipment and tooling costs. In this connection the automotive and appliances industries are the major users of moulded rubber parts in large volume including such items as sealing grommets, moulded hoses, ball joint dust boots, bellows for rack and pinion stearing and plug connectors. Neoprene, Hypalon and Viton, are frequently used in the manufacture of these parts where service conditions demand good resistance to lubricating oils, fuels, greases, chemicals, heat and ozone. Undoubtedly these elastomers, particularly Neoprene, will therefore find increasing application in injection moulding, particularly where articles are thin walled with limited volume, thus offering the possibility of rapid cure rate, and a high degree of dimensional accuracy and good surface finish not normally obtainable on compression moulding without excessive rejects.

13.2 MACHINE OPERATING CONDITIONS

With both types of machine there are a number of operating conditions which must be correctly adjusted, depending on the compound and cavity volume, to obtain good parts in the shortest possible total cycle time. Table 13.1 gives these conditions and the manner in which they are controlled in the two basic types of machine.

The effect of machine conditions on injection time, cure rate and cycle time may now be considered.

13.2.1 RESERVE STOCK TEMPERATURE

For the shortest cure cycle the cylinder or barrel temperature should be as high as possible without scorch or blistering occuring. High cylinder or barrel temperature tends to lower stock viscosity and thus shorten injection time. Barrel temperature is controlled by liquid circulation; water at 70° to 95°C being the most common heat transfer medium at present. Temperatures above 95°C may be required for very stiff stocks and can be obtained by using glycerine in the jacket.

Table 13.1

Condition	Controlled by	
	(In ram machine)	(In screw machine)
Reserve stock temperature	Cylinder temperature	Barrel temperature
Heat build-up in orifice	Nozzle diameter and injection pressure	Nozzle diameter and injection pressure
Mould temperature	Platen temperature Control	Mould temperature controls*
Injection pressure	Ram pressure	Screw pressure
Clamp time	Platen clamp cycle	Mould clamp cycle*

*   Heated platens can be fitted to the screw type machine.

Temperatures below 70°C may be desirable on very soft compounds to prevent blisters and air trapping. The cylinder or barrel temperature affects the rate of cure, since it directly controls stock temperature as it enters the mould. Increasing reserve stock temperature is frequently more effective in shortening cure cycle times than increased acceleration in the compound. The effect of reserve stock temperature on cure rate is shown in Table 13.2; the compression set results demonstrate the tighter cure obtained with higher cylinder temperature. The scorch characteristics of a compound determine the safe upper limit of the reserve stock temperature and its total dwell time in the cylinder or barrel.

Table 13.2 Effect of cylinder temperature on rate of cure of a Neoprene compound

Ram type machine
Injection pressure (1 758 kg./sq. cm.)
Clamp time—2½ min. at 204°C

Cylinder temperature °C	82	93	104
Tensile strength, kg./sq. cm.	98	102	109
Elongation at break, %	500	450	375
Hardness, Durometer A	64	64	65
Tear strength (ASTM D624, die C), kg./sq. cm.	57	50	47
Compression set, method B, %, 22 hr. at 70°C.	21	17	9

13.2.2 NOZZLE DIAMETER

Three factors require balancing in determining optimum nozzle diameter:

1 The heat increase through the nozzle, which is a function of the nozzle diameter and of available injection pressure.

2 The pressure required to inject the stock, which depends on the compound viscosity and the base elastomer used.

3 The time taken for the actual mould filling operation, which should be held to a minimum to avoid precuring during filling.

With the reciprocating screw type machine which is under consideration, the smallest nozzle size consistent with reasonable injection time is about 2½ mm. diameter; this results in a temperature rise as the stock passes through the nozzle of the order 28° to 56°C depending on the elastomer which is being processed and the compound viscosity.

A further consideration is the size of the cavities which are to be swept by the stock. If the compound cures up in the mould before the cavity is full, the nozzle is obviously too small. Another factor in determining optimum nozzle diameter is the design of the mould runners and gates. A high temperature rise can result by feeding through very small diameter gates, i.e. 0·1 mm.

13.2.3 MOULD TEMPERATURE

In ram type machines the temperature of the platen affects injection time, since it alters the viscosity of the compound during flow through runners and gates, thus injection time is reduced by increasing platen and mould temperatures unless precure of the stock occurs.

The reciprocating screw machine as manufactured may or may not utilize heating platens. Moulds with deep draw used on these machines are provided with built in resistance heating elements located at strategic points to ensure balanced heat transfer to the compound. Alternatively, if the moulding is shallow or if the stock has to run some distance through runners on an outside face of the mould, heated platens can be fitted to the screw type machines.

Electrically heated platens or built in resistance type elements are invariably the source of heat in injection moulding equipment. Temperatures varying from 182° to 217°C are currently used for curing Neoprene compounds in production.

13.2.4 INJECTION PRESSURE

Injection pressure directly affects injection time; the higher the pressure the shorter the injection time. Compound viscosity and the basic polymer used are the factors in determining injection pressure. Injection pressure must also be related to clamping force. Too high an injection pressure coupled with insufficient clamping force will result in the mould opening and flashing occuring. Low viscosity stocks may be injected faster than the air can escape from the mould, in which case reduced injection pressure is essential.

13.2.5 CLAMP TIME

This is the interval between closing and opening of the mould and includes injection time, hold time and cure time. Hold time is a short period after injection ceases when much reduced pressure is maintained to consolidate the moulding during the initial stages of cure, thus preventing any escape of stock from the mould. Total cycle time is the overall time between successive closings with the mould.

Clamp time and cycle time both affect injection time, particularly in ram type machines because the longer the stock remains in the cylinder before injection, the warmer it becomes and the faster it will flow under the available injection pressure.

Table 13.3 shows the effect of clamp time on the cure state of a Neoprene compound.

Table 13.3 Effect of clamp time on state of cure of a Neoprene compound

Ram type machine
Injection pressure (1 758 kg./sq. cm.)
Cylinder temperature (82°C)
Mould temperature (218°C)

	1½	1¾	2
Clamp time, min.			
Tensile strength, kg./sq. cm.	106	106	116
Elongation at break, %	420	310	325
Hardness, Durometer A	67	68	67
Tear strength, (ASTM D624, die C), kg./sq. cm.	56	55	50
Compression set, method B, %, 22 hr. at 70°C	34	18	13

In reciprocating screw machines there is less need for heat transfer from barrel to stock because of the heat produced by the rotating screw. Clamp time and total cycle time have, therefore, less effect on injection time.

13.2.6 FORM OF FEED STOCK

High viscosity compounds (highly loaded or based on crystallizing polymers) may cause feeding difficulties in screw type machines. Granular or chipped feed is preferred with these stocks to ensure that sufficient stock is available at the head of the screw for the injection stroke. In ram type machines the feed stock is usually extruded but its form has no bearing on injection pressure or flow.

13.3 DU PONT ELASTOMERS IN INJECTION MOULDING (GENERAL)

In many cases a compression moulding compound can also be injection moulded through the proper choice of machine settings; the exceptions are high viscosity and slow curing formulations.

Low viscosity formulations, will inject more rapidly than high viscosity compounds. A desirable range of Mooney viscosity values would be in the range of 15–80 (ML 100°C, 2½ min.).

Slow curing formulations requiring 30 min. or more at 153°C for optimum cure will require uneconomically long cure cycles. Any compound with a Mooney scorch of 15–30 min. for a 10 point rise at 121°C should injection mould satisfactorily. 'Scorchier' compounds can be handled if special precautions are taken.

The state of cure of an injection moulded part where the cross-section is thin, 2·5 mm. or less, is only a little better than that of the same part cured the same length of time at the same temperature in a compression mould. In thicker parts injection moulding develops a higher state of cure. This is due to the more uniform temperature throughout the part at the start of the cure.

In common with other elastomers, closer control of compound viscosity and scorch resistance is required when Neoprene, Hypalon or Viton compounds are injection moulded.

Owing to the greatly increased productivity obtained from the moulding tool, some increase in mould cleaning may be necessary. However, the type of steel used in the construction of the mould is important to minimize this condition, nickel-chrome or stainless steels being preferred. It may also be necessary to increase the level of the acid acceptor in the compound.

The following mould release formulation is recommended for use in injection moulding procedure:

    0·5 part DFL No. 3 (buffered phosphate ester)
    0·5 part silicone emulsion (35%)
    99 parts distilled or de-ionized water.

It is important when preparing DFL No. 3 solutions that distilled or de-ionized water is used to prevent the formation of calcium salts which tend to plug the ends of spray hoses and nozzles. The use of DFL No. 3 also inhibits corrosion of mould surfaces during storage.

13.4 NEOPRENE

The W Neoprenes are preferred for injection moulding, since they provide consistent compound viscosity from one batch to the next and permit the adjustment of cure rate and scorch resistance, by means of various accelerator systems. The W Neoprenes offer a comprehensive viscosity range both in crystallizing and non-crystallizing grades (Table 13.4), thus enabling optimum compound viscosity for maximum

Table 13.4 Neoprene W types

	Mooney viscosity (ML 100°C–2½ min.)	Crystallization rate
Neoprene WM-1	36–42	Fast
Neoprene W	45–54	Fast
Neoprene WHV	110–130	Fast
Neoprene WRT	45–54	Very slow
Neoprene WX	45–54	Medium
Neoprene WD	110–130	Very slow

trouble free production to be readily obtained, particularly by the use of the blends.

If a G type Neoprene is essential for technical purposes in the product such as optimum tear strength or flex resistance, Neoprene GT should be considered, since this new G type polymer can be readily peptized to the desired viscosity and cure rate can also be adjusted with NA-22. The following factors are pertinent to maximum processing safety on

Neoprene compounds, particularly with regard to freedom from set up in the barrel or cylinder of the injection machine.

First, always use a grade of magnesia which possesses high activity in order to obtain maximum scorch protection and maintain the condition of this magnesia excluding moisture during storage by purchasing it in polyethylene prepacks.

Secondly, select the antioxidant with care; certain classes of antioxidant adversely affect the processing safety of Neoprene compounds, in particular ketone amine and quinoline classes should be avoided. When paraphenylene diamine derivatives are necessary for additional protection against ozone cracking, the effect on scorch and bin storage must be taken into account.

Thirdly, several accelerator systems giving a good balance of rate of cure and processing safety are available for W type compounds. The use of NA-22 in conjunction with TMTDS or MBTS is of particular interest, since it gives faster cure rate and improved scorch resistance compared to NA-22 alone. The TMTMS, DOTG, sulphur system gives maximum scorch protection and may be preferred on hot running compounds such as high hardness, high viscosity stocks. The cure rate of this system can be increased by the addition of NA-22 as the secondary accelerator, to obtain very fast cycle times with adequate scorch resistance. A typical suggestion is DOTG 0·75, TMTMS 0·75, sulphur 0·5, NA-22 0·25. Stearic acid should be limited to a maximum 0·5 parts in Neoprene W compounds, since it is not as compatible as in natural rubber. To minimize mould dirtying, up to 8 phr of magnesia can be used in the compound. Mineral fillers tend to cause porosity and should be avoided. For the same reason process oils with high volatility are preferred.

Tables 13.5 and 13.6 illustrate results obtained with low and medium viscosity Neoprene compounds on a reciprocating screw machine. Injection required 3–5 sec.; and a 45–60 sec. cycle at 204°C gave a state of cure equivalent to that of compression moulded test slabs cured 20 min. at 153°C.

High viscosity compounds giving rapid build-up in nozzle temperature due to frictional heat may require modification by the introduction of 10–20 parts of fluid polymer, Neoprene FC. This is an effective method of lowering compound viscosity whilst maintaining high cured hardness, since the fluid polymer vulcanizes. The use of internal lubricants in a high viscosity compound is a further means of increasing injection rate if this is required. The use of IML-1 also gives lower mixing temperatures and is a means of lowering modulus where extraction problems are encountered on undercut sections.

Table 13.7 gives the results obtained with a modified very high viscosity compound on the reciprocating screw machine. The introduction of 10 parts Neoprene FC and 2·5 parts IML-1 in conjunction with chip feed was necessary to fill the cavities at the maximum injection pressure of 1125 kg./sq. cm. With this compound a 35 sec. cycle at 204°C gave a satisfactory state of cure.

13.5 HYPALON, CHLOROSULPHONATED POLYETHYLENE

Hypalon 40 and 45 are easy processing polymers better suited for

Table 13.5 Low viscosity Neoprene compound 50 Durometer A hardness

Neoprene WRT	100
Magnesia	4
Neozone A	2
Heliozone	1
Stearic acid	0·5
MT Carbon black	25
Light process oil	20
Zinc oxide	5
NA-22	1

Mooney scorch—MS 121°C	
Min. to 10 point rise	16·5

Mooney Viscosity—ML 100°C	20·5
2½ min. value	

MACHINE CONDITIONS (Edgwick 45-SR)

Barrel temperature	82°C
Mould temperature	204°C
Screw speed	50 rpm
Injection pressure	471 kg./sq.cm.
Hold-on pressure	352 kg./sq.cm.
Injection nozzle diameter	2·4 mm.
Feed	Strip
Cycle times	As shown below

PHYSICAL PROPERTIES*

Injection moulded at 204°C (Cycle time includes 3 sec. injection time)

Cycle time, sec.	45	60	75	90	105
Tensile strength, kg./sq.cm.	107	109	109	107	104
Elongation at break, %	630	610	610	610	600
Modulus at 300%, kg./sq.cm.	19	26	30	30	30
Hardness, Durometer A	46	49	49	48	50

PRESS CURED AT 153°C

Min.	10	20	30	40
Tensile strength, kg./sq.cm.	72	111	113	104
Elongation at break, %	900	650	630	580
Modulus at 300%, kg./sq.cm.	18	25	28	28
Hardness, Durometer A	42	48	49	50

* Moulded part shaped like a flower pot 3 in. high with diameter 4·5 in. at open end, 3 in. at closed end, and 0·08 in. wall; gate in 3-in. end.

187

injection moulding compounds than Hypalon 20 and 30. Hypalon 40 having a Mooney viscosity at 100°C of 60 ±6 compared to Hypalon 45 with 37 ±6. Intermediate viscosities can be obtained by the use of blends. Several curing systems for Hypalon compounds are available based on either litharge, magnesia/pentaerythritol or epoxy resins. Tetrone A (di pentamethylene thiuram tetrasulphide) and MBTS are the principle sulphur bearing accelerators used. Stearic acid should not be used with

Table 13.6 Medium viscosity Neoprene compound 65 Durometer A hardness

Neoprene W	100
Neozone A	2
Stearic acid	0·5
Petrolatum	1
Magnesia	4
MT Carbon black	100
Light process oil	10
Zinc oxide	5
NA-22	0·5

Mooney scorch–MS 121°C

Min. to 10 point rise	12·5

Mooney viscosity–ML 100°C

2½ min. value	64

MACHINE CONDITIONS (Edgwick 45-SR)

Barrel temperature	71°C
Mould temperature	204°C
Screw speed	50 rpm
Injection pressure	471 kg./sq.cm.
Hold-on pressure	352 kg./sq.cm.
Injection nozzle diameter	2·4 mm.
Feed	Strip
Cycle times	As shown below

PHYSICAL PROPERTIES*
Injection moulded at 204°C (Cycle time includes 5 sec. injection time)

Cycle time, sec.	30	45	60	75	90	105
Tensile strength, kg./sq.cm.	128	130	130	132	135	137
Elongation at break, %	430	420	420	420	410	390
Modulus at 300%, kg./sq.cm.	98	109	111	113	116	116
Hardness, Durometer A	63	66	67	68	68	70

PRESS CURED AT 153°C

Min.	10	20	30	40
Tensile strength, kg./sq.cm.	120	134	135	137
Elongation at break, %	530	430	390	390
Modulus at 300%, kg./sq.cm.	79	109	116	121
Hardness, Durometer A	60	65	67	68

* Moulded part shaped like a flower pot 3 in. high with diameter 4·5 in. at open end, 3 in. at closed end, and 0·08 in. wall; gate in 3 in. end.

Table 13.7 High viscosity Neoprene compound 90 Durometer A hardness

Neoprene WRT	90
Neoprene FC	10
Neozone A	2
Magnesia	4
Stearic acid	0 5
MT Carbon black	60
SRF Carbon black	100
DOTG	0·5
Thionex	0·5
Zinc oxide	5
Sulphur	1
Light process oil	10
IML-1	2·5

Mooney scorch—MS 121°C
Min. to 10 point rise — 27

Mooney viscosity—ML 100°C
2½ min. value — 164

MACHINE CONDITIONS (Edgwick 45-SR)

Barrel temperature	71°C
Mould temperature	204°C
Screw speed	50 rpm
Injection pressure	1 153 kg./sq.cm.
Hold-on pressure	373 kg./sq.cm.
Injection nozzle diameter	2·4 mm.
Feed	Granules
Cycle times	As shown below

PHYSICAL PROPERTIES*
Injection moulded at 204°C (Cycle time includes 5 sec. injection time)

Cycle time, sec.	35	45	60	75	90
Tensile strength, kg./sq.cm.	135	134	134	132	132
Elongation at break, %	140	120	110	110	110
Hardness, Durometer A	86	87	87	90	92

PRESS CURED AT 153°C

Min.		5	10	20	30
Tensile strength, kg./sq.cm.		83	118	132	132
Elongation at break, %		240	180	130	110
Hardness, Durometer A		80	90	90	91

* Moulded part shaped like a flowerpot 3 in. high with diameter 4·5 in. at open end, 3 in. at closed end, and 0·08 in. wall; gate in 3 in. end.

litharge based curing systems, since it causes rapid set up of mixed stock. Two-stage mixing of Hypalon compounds is normally recommended, the acclerators being added on the second pass.

Table 13.8 gives the results obtained on a reciprocating screw machine

Table 13.8 White Hypalon for moulded goods 70 Durometer A hardness

Hypalon 40	100
Titanium dioxide	20
Hard clay	25
Whiting	50
Di octyl phthalate	15
Pentaerythritol (200 mesh)	3
Tetrone A	2
Non-discolouring antioxidant	1
Magnesia	4

Mooney scorch—MS 121°C
Min. to 10 point rise   31

Mooney viscosity—ML 100°C   60
(2½ min. value)

MACHINE CONDITIONS (Edgwick 45-SR)
Barrel temperature   71°C
Mould temperature   204°C
Screw speed   50 rpm
Injection pressure   470 kg./sq.cm.
Hold-on pressure   352 kg./sq.cm.
Injection nozzle diameter   2·4 mm.
Feed   Strip
Cycle times   As shown below

PHYSICAL PROPERTIES*
Injection moulded at 204°C (Cycle time includes 3 sec. injection time)

Cycle time, sec.	*60*	*75*	*90*
Tensile strength, kg./sq.cm.	127	134	130
Elongation at break, %	580	570	560
Modulus at 300%, kg./sq.cm.	49	51	53
Hardness, Durometer A	60	64	65

PRESS CURED AT 153°C

Min.	*10*	*20*	*30*	*40*
Tensile strength, kg./sq.cm.	120	132	134	135
Elongation at break, %	600	540	540	520
Modulus at 300%, kg./sq.cm.	49	53	55	56
Hardness, Durometer A	65	66	67	68

* Moulded part shaped like a flower pot 3 in. high with diameter 4·5 in. at open end, 3 in. at closed end, and 0·08 in. wall; gate in 3 in. end.

with a 70 Durometer A white Hypalon compound with magnesia/ pentaerythritol curing system, accelerated with Tetrone A. The compound had medium viscosity; no difficulty was experienced with injection. The 90 sec. cure at 204°C gave a state of cure equivalent to that of compression moulded slabs cured 20 min. at 153°C.

Table 13.9 illustrates a 70 Durometer A black Hypalon compound with an epoxy resin curing system with Tetrone A as the primary accelerator. The compound was fully cured with a 50–60 sec. cycle at 204°C.

13.6 VITON FLUOROELASTOMER

Viton A, B, and A/HV compounds are usually higher in viscosity than

Table 13.9 Black Hypalon for moulded goods 70 Durometer A hardness

Hypalon 40	100
Hard clay	50
MBTS	0·5
Tetrone A	1·25
Epoxy resin	15
DOTG	0·25
FT carbon black	25
Liquid coumarone-indene resin	10
Low molecular weight polyethylene	4

Mooney scorch—MS 121°C

Min. to 10 point rise	25

Mooney viscosity—ML 100°C

2½ min. value	56

MACHINE CONDITIONS (Edgwick 45-SR)

Barrel temperature	71°C
Mould temperature	204°C
Injection pressure	372 kg./sq.cm.
Hold-on pressure	186 kg./sq.cm.
Injection nozzle diameter	2·4 mm.
Feed	Strip
Cycle times	As shown below

PHYSICAL PROPERTIES*
Injection moulded at 204°C (Cycle time includes 3 sec. injection time).

Cycle time, sec.	45	50	60	75	90
Tensile strength, kg./sq.cm.	158	164	178	172	169
Elongation at break, %	610	640	640	640	640
Modulus at 300%, kg./sq.cm.	49	53	51	49	49
Hardness, Durometer A	64	65	64	67	66

PRESS CURED AT 153°C

Min.	5	10	20	30
Tensile strength, kg./sq.cm.	127	148	149	151
Elongation at break, %	830	730	670	660
Modulus at 300%, kg./sq.cm.	35	46	55	56
Hardness, Durometer A	60	66	68	69

* Epikote Resin 828 was used.
  Moulded part shaped like a flower pot 3 in. high with diameter 4·5 in. at open end, 3 in. at closed end, and 0·08 in. wall; gate in 3 in. end.

compounds of other polymers having the same vulcanizate hardness.
Viton A35 is recommended for Viton injection moulding compounds,
since this new type has lower Mooney viscosity.

The moulds for Viton should be highly polished and chrome plated.
A mould lubricant should be sprayed into the cavities before each cycle,
to improve release of the cured parts. Table 13.10 shows the physical
properties obtained on injection moulding an 80 Durometer
Viton compound.

Table 13.10 80 Durometer A Viton compound

Viton A35	100
Low activity magnesia	15
MT Carbon black	25
DIAK No. 3	2·5

Cured at 232°C in Lewis injection press for:

	2 min.		3 min.	
Oven post cure at 204°C	None	24 Hr.	None	24 Hr.
ORIGINAL PHYSICAL PROPERTIES				
100% Modulus, kg./sq.cm.	28	42	30	49
200% Modulus, kg./sq.cm.	60	97	60	104
Tensile strength, kg./sq.cm.	118	155	102	111
Elongation at break, %	500	340	340	210
Hardness, Durometer A	77	78	76	79
COMPRESSION SET				
Method B, %,				
70 hr. at 149°C	79	41	–	–

Pieces were round slabs, 13·3 cm. in diameter and 2·5 mm. thick. Cavity filled in 8 sec. with
91 kg./sq. cm. injection gauge pressure. Cylinder temperature was 82°C.

Du Pont registered trade marks

Heliozone	sun checking inhibitor
Hypalon	synthetic rubber
Neozone	rubber antioxidant
Tetrone	rubber accelerator
Thionex	rubber accelerator
Viton	fluoroelastomer

Du Pont registered trade mark

Diak	curing agent
	(for Viton fluoroelastomer)

13.7 TROUBLE SHOOTING GUIDE

DISTORTION

Distortion on removal from the mould may be caused by too long an injection time. To remedy this problem:

(a) Increase injection pressure.

(b) Use fresh stock. Old stock may have precured.

(c) Recompound for lower viscosity and better flow.

(d) Check and perhaps enlarge nozzle, runners and gates. (Note: This step is irrevocable. Gates should be enlarged only as a last resort.)

ORANGE PEELING

This may be caused by precure occuring during the filling of the mould cavity. To remedy this problem:

(a) Reduce mould temperature.

(b) Adjust accelerator system to increase processing safety and reduce cure rate.

LONG CURE CYCLES

This may be caused by low mould or reserve stock temperatures or an inadequate cure rate. To remedy this problem:

(a) Increase mould or reserve stock temperature individually or together.

(b) Adjust accelerator to obtain increased cure rate.

(c) Increase compound viscosity by changing polymer or filler, thus obtaining increased friction and temperature rise at the nozzle and the gates.

BLISTERS

Blisters, due to trapped or entrained air. To remedy this problem:

(a) Decrease injection pressure to increase injection time. The slower injection rate will reduce the probability of entrainment of air as the stock shoots into the empty passages in the mould.

(b) Reduce temperature of the reserve stock, thus increasing stock viscosity.

(c) Compound for increased stock viscosity.

(d) Vent the mould properly and adequately.

Blisters caused by entrappment of gases other than air. To remedy this problem:

(a) Check all compounding ingredients for volatility at curing temperatures. Look for moisture in the compound.

(b) Decrease temperature of the reserve stock.

(c) Decrease mould temperature.

BACKRINDING

Backrinding or 'sink-back' at the gates may be caused by the accelerator system or by the hold-on time that is too long. To remedy this problem:

(a) Decrease the level of the fastest curing accelerator in the curing system.

(b) Decrease temperature of the mould.

(c) Decrease 'hold-on time'

GRAIN EFFECT

This is exhibited by poor tear strength in one direction and is caused by the stock not being thoroughly mixed in the mould before the beginning o of the cure cycle. This effect does not occur with reciprocating screw type machines. To remedy this problem:

(a) Increase the temperature of the reserve stock and mould, thus achieving better mixing after the stock leaves the gate.

(b) Increase turbulence in the sprues by use of a smaller nozzle.

Note: Grain effect should not be confused with 'orange peeling'. Remedies for one will only aggravate the other.

SURFACE CONTAMINATION

This is due to mould lubricants or decomposition of the moulded article. To remedy this problem:

(a) Increase the amount of stabilizer in the compound (e.g. metal oxide in Neoprene).

(b) Use a good mould lubricant.

AIR TRAP

Any air trapped in the mould cavity will prevent proper filling. To avoid trapping air:

(a) Design the mould so that there are no obstacles to prevent expulsion of air from all parts of the mould as the stock fills the cavities.

(b) Eliminate all sharp 'dead-end' corners in either the cavity or the overflow groove.

(c) Increase injection time by increasing stock viscosity or by lowering the injection pressure.

13.8 CONCLUSIONS

In conclusion the author would like to thank the organizers of this Conference on behalf of the Du Pont Company for inviting him to present this paper. It is hoped that it has illustrated that Neoprene, Hypalon and Viton compounds can be injection moulded on both the two basic types of equipment now available to produce better mouldings at lower unit cost provided machine operating conditions are correctly adjusted and the compounding and processing of the elastomer is carefully controlled.

The broad range of Du Pont elastomers available provides the compounder with many degrees of freedom in developing suitable compounds for injection moulding to meet a variety of end-use applications. The proper control of viscosity, scorch and cure rate to fit the type of equipment in use appears to be the most important factor in compounding for satisfactory injection moulding. For the future the author is convinced that injection moulding is going to revolutionize the moulded goods field in the next few years. However, full economics can only be realized in the injection moulding process with long runs with the same product or at least the same formulation. The number of compound variables will therefore have to be drastically reduced for this to be practical. There will be fewer elastomeric compounds, with a premium on stocks which will do a variety of jobs to the point where a composition which can meet a variety of requirements will command a relatively high raw materials cost, because of inventory and production economies it can offer.

REFERENCES

1 Injection Moulding Compounding and Equipment—*Rubber World,* July 1963.

2 Gregory, C. H., Injection Moulding of Rubber Mechanicals— A pause for reflection—*Rubber Journal,* August 1964.

3 Gregory, C. H., Injection Moulding of Rubber Goods—A review of the current position—*Rubber Journal,* October 1964.

DISCUSSION

W. S. PENN *(Borough Polytechnic)*: Could you please comment on mould fouling?

ANSWER: Mould fouling is a major problem in injection moulding, since it can adversely affect the economic advantages which injection moulding offers.

In our experience with Neoprene, we have found that the tool steel is of major importance in overcoming this problem. Nickel-chrome steels are preferred for injection moulding tools. In the event of mild steel tools being used and excessive fouling being experienced, considerable relief may be obtained by increasing the level of acid acceptance in the Neoprene compound, i.e. increasing the magnesia level from 4 parts to 8 parts has improved mould cleaning from once in every 24 hours to once per week.

H. M. GARDNER *(T. H. & J. Daniels Ltd)*: Please comment on air trapping.

A: The author agrees that air trapping in the mould may at times be more easily eliminated by directly reducing injection speed rather than injection pressure. The aim should be to slow the rate of injection into the cavity, so as to allow residual air in the cavity to escape through the venting points.

T. SEKLECKI *(GKN Machinery Peco Division)*: Is special precaution necessary in selection of mould material to prevent excessive fouling when moulding soft nitrile compounds?

A: We believe special precaution is necessary in the selection of tool steels to prevent excessive fouling with soft nitrile compounds. Nickel-chrome steels are preferred for injection moulding tools.

P. BOIS *(Industrial Plastics Ltd)*: Hose and bellows—Do these products necessarily have to be produced on very expensive multi-station machines?

A: We believe it is essential to consider multi-station machines for the large scale production of injection moulded bellows.

The use of multi-station machines eliminates change time as a factor in the cure cycle.

The author agrees that it is possible to injection mould long length hoses using interchangeable cores. Iti si doubtful whether this would be a proposition for the very wide range of moulded bellows which are required by the automotive industry.

CHAPTER 14

Injection moulding of silicone rubbers

H. W. WINNAN, A.N.C.R.T., A.I.R.I.
Midland Silicones Ltd

14.1 INTRODUCTION

SILICONE rubbers were first made commercially available some twenty-five years ago, since which time they have become an established part of the rubber industry's complex make up. Those first compounds which seemed to require so much care to vulcanize and to keep free from contamination have been replaced by materials which are readily accepted where their specialized properties are required. The range available today is enormous—everything from translucent to conductive, from low temperature to solvent resistant, from premium cable insulating grades to no-post cure stocks. The emphasis in the silicone rubber market has always been on fully compounded materials which essentially just need shaping and vulcanization, and broadly speaking, as new requirements become apparent so new grades are developed to suit.

It is rather interesting that although injection moulding offers many advantages both technically and economically in certain cases, there has been no apparent extending of the range of grades to include one specifically for injection moulding.

It is not the concern of this short chapter to set out to analyse the conditions which are right for considering making items by injection moulding as against any other technique. Rather, it is proposed to pass on the information gained during some preliminary trials with various grades of Silastomer and to explain some quite new developments which are allied to the injection moulding principle.

14.2 EXPERIMENTAL DETAILS

Five grades of silicone rubber were examined on the Edgwick 45 SR machine which is of the screw ram type and was made available for this study by the courtesy of T. H. and J. Daniels Ltd, of Stroud, to whom the author is indebted.

The feed arrangement on this particular model was a simple throat having an undercut section adjacent to the screw, but *not* fitted with a roller feed as is normally preferred on extruders for silicone rubber.

The L/D ratio of the screw was 12:1 and the compression ratio 1·5:1. Temperature control of the barrel was maintained by circulating water heated by thermostatically controlled immersion heaters in a separate unit.

A single cavity mould was used which was electrically heated and produced a flat disc of approximately 4 in. diameter and thickness $\frac{1}{10}$ in. The nozzle diameter used throughout the evaluation was $\frac{3}{32}$ in.

The five grades of silicone rubber examined were as follows:

1 A nominal 50° B.S. hardness no-post cure grade of Williams plasticity 220 and Mooney viscosity (measured at room temperature) 30. The term no-post cure is appropriate here since unlike conventional grades of silicone rubber, this grade attains its optimum properties on moulding without the necessity for a post cure. Hardness and compression set are the properties of conventional grades of silicone rubber most susceptible to change during the four to 24 hr. post cure at 200–250°C normally given to grades of silicone rubber.

2 An experimental silicone rubber grade having the unusually high Mooney viscosity of 100, and Willaims plasticity of 1 000. This material is of nominal 90°B.S. hardness after a four-hr. 200°C post cure. The appearance and nature of this material both cured and uncured is quite untypical of silicone rubbers generally.

3 A general purpose/moulding/extrusion/calendering grade of nominal hardness 60 B.S. after a post cure and of Mooney 40–45 and Williams plasticity 250.

4 A high strength grade of nominal hardness 50° B.S. and Mooney 60, Williams plasticity 400.

5 An extremely low temperature grade suitable only for moulding purposes, of nominal hardness 50° B.S. Mooney viscosity 30 and Williams plasticity 210.

Each of these grades was milled for 5 min. to break down any structure which may have formed owing to the interaction of reinforcing silicas with siloxane gums. The milled sheet of approximate thickness ¼ in. was cut into 1 in. wide strips and fed into the throat of the machine by hand.

The barrel temperature was set at between 48° and 100°C depending on the peroxide present in the compound.

The injection time, injection pressure, holding pressure, time under pressure, moulding time and temperature were then varied until satisfactory mouldings were produced. The moulding time was then gradually reduced, keeping constant as many of the other variables as possible until the minimum moulding time for a grade was reached.

The change from one compound to another was by the conventional purging technique which was found to require approximately 2 lb. of compound in this particular machine to remove all but the last traces of the previous compound.

Type E dumb-bell test pieces were cut from the discs produced and physical properties of 'as moulded' and post cured specimens were compared with conventional compression moulded specimens from the same batch of each grade examined.

14.3 RESULTS

The following are some of the conclusions reached.

With the commonly used 2:4 dichlorobenzoyl peroxide, the maximum safe barrel temperature is 50°C. This compares with at least 100°C for dicumyl peroxide-containing compounds, this temperature being the maximum obtainable since water was being used as the circulating liquid.

Moulding temperatures only slightly higher than 115°C normally recommended for the dichlorbenzoyl peroxide were possible, whereas with less active peroxides like dicumyl peroxide, mould temperatures in excess of 200°C were possible instead of the 150°C normally used for this peroxide.

Moulding times of 30 sec. or less were possible for all of the compounds examined except the no-post cure grade containing dicumyl peroxide where 60 sec. at 170°C was required, compared with a minimum of 300 sec. under compression moulding conditions.

Injection pressures varied from 6 850 psi for the 100 Mooney compound down to as low as 2 050 psi for the 30 Mooney compound. In both cases 2 sec. or less was a suitable injection time with the standard $\frac{3}{32}$ in. diameter nozzle. The pressures used are well within the capability of the machine which has an upper limit normally of 16 400 psi.

Once the rubber has been injected into the mould and held under a given pressure for the required time, the pressure is allowed to drop off to a lower value called the holding pressure. For most of the grades examined, the holding pressure could be dropped to approximately half the injection pressure after a few seconds.

The processing characteristics of all of the grades were good as gauged by the ease of feeding the strip into the machine although not in the same order of processability for injection moulding as for compression moulding. For instance, the no-post cure grade which has excellent processing properties normally, was found to give some difficulty owing to the problem of maintaining a continuous feed at the throat of the machine. Equally, the uncharacteristically high Mooney grade which normally causes difficult conventional processing, was found to be the best processing grade of all during injection moulding.

The comparison of vulcanizate properties with compression moulded sheets showed only marginal differences between the two. Tensile and tear strength were generally higher whereas hardness and elongation were lower in most cases. Shrinkage was higher with injection moulded samples but this is due to higher moulding temperatures on injection moulding. Since silicone rubbers have a high coefficient of thermal expansion (7×10^{-4} average, equivalent to approximately 100 times that of mild steel) this is to be anticipated.

The hot air ageing characteristics of injection mouided material are marginally better than the compression moulded counterparts, as measured by retention of tensile strength. Hardness and elongation changes are not significantly affected.

14.4 MORE RECENT WORK

There have been several developments affecting the processing of silicone rubbers which have taken place since this work was carried out. The innovation of non milling stocks is one. With this type of material, so far developed for cable grades mainly, the form is that of a talced solid circular rod of approximately 1 in. diameter. A convenient container may hold 50 lb. of rubber in a continuous length of talced rod so that once it has started to feed into the machine only the very minimum of attention is required by an operator to ensure continuity of feed.

The development of non milling stocks has also led to the introduction of pelletized silicone rubbers. These pellets are talced to prevent agglomeration and may be charged to the hopper of an Edgwick 45 SR machine which holds approximately 25 lb. of rubber. The pelletized rubber is automatically vibrated and trials on an extruder show that the system is practicable with silicone rubber. Where coloured mouldings are required, the addition of approximately 10% of pelletized colour master batches suitably mixed with the rubber pellets has been found to produce uniformly pigmented extrusions for instance. Although we have not had an opportunity to try out this technique in the 45 SR machine there appears to be no reason why such a simple system cannot be adopted with reasonable success.

There is no reason incidentally why silicone rubber in pelletized form should be restricted to non milling grades. If the grade is adequately plasticized by the screw of an extruder or injection moulding machine, then conventional milling grades can be injection moulded, coloured or otherwise, by this simple technique.

14.5 CONCLUSIONS

In conclusion, it is apparent that silicone rubber as we know it today is eminently suitable for the injection moulding process. The use of the less reactive peroxides such as dicumyl peroxide gives greater freedom from scorching although this is not considered to be a problem since there is little heat build-up with these inherently low viscosity materials. Although silicone rubbers do not show such a marked viscosity/temperature relationship as do organic rubbers, raising the temperature at the point of injection from say $50°$ to $100°C$ will be beneficial.

In principle one wants to aim at the minimum temperature rise between barrel and mould, consistent with scorch free processing.

The reduction in moulding time claimed by other workers in this field, has been established with only marginal differences in vulcanizate properties.

A reduction in the peroxide level for injection moulding as compared with compression moulding is also claimed by other workers, with consequent improvements in compression set. This is one area where there is a considerable amount of work still to be carried out.

It is proposed to undertake further detailed studies into the process.

Note: Silastomer is the registered trade mark of Midland Silicones Ltd.

DISCUSSION

W. S. PENN *(Borough Polytechnic)*: How is the problem of expensive sprue and runners dealt with in the case of silicone rubbers?

Is injection moulding being used commercially with silicones?

ANSWER: Although I have not carried out any experiments with reclaimed scrap silicone rubber for injection moulding purposes, it is normally possible to use up to 30% of such material milled into virgin material for moulding purposes. It is important that scrap material is only blended back into fresh material of the same grade to ensure the best results and extra peroxide must be added to the blend to compensate for the reclaim fraction since this will not contain any curing agent.

On the question of the commercial use of silicone rubbers for injection moulding, I know only of two companies using this process, but the information is not easy to obtain for an informed answer.

Dr L. J. GERHARDT *(Vitamol Precision)*: What can be done to improve tear strength of silicone rubber as the moulding temperature is higher with less active peroxides?

A: The tear strength of silicone rubber has been improved for some of the more recently released grades. However most of these use a very active peroxide like 2:4 dichlorobenzoyl peroxide. With no post-cure grades, which need the vinyl specific type of peroxide, the hot tear strength has not been found to cause any additional problems and it has been shown that some improvement may be gained by the choice of secondary filler in such compositions.

K. J. TURK *(T. H. and J. Daniels Ltd)*: Would a spiral flow test mould similar to that used for plastics be a suitable means for evaluating rubber flow properties, i.e. pressures, *versus* temperatures, etc?

A: Although I do not feel really qualified to answer this question, I think it would be a useful tool for such studies.

Index